Corrosion Control Through Organic Coatings

CORROSION TECHNOLOGY

Editor
Philip A. Schweitzer, P.E.
Consultant
York, Pennsylvania

Corrosion Control Through Organic Coatings

Amy Forsgren

Taylor & Francis
Taylor & Francis Group
Boca Raton   London   New York

A CRC title, part of the Taylor & Francis imprint, a member of the
Taylor & Francis Group, the academic division of T&F Informa plc.

Published in 2006 by
CRC Press
Taylor & Francis Group
6000 Broken Sound Parkway NW, Suite 300
Boca Raton, FL 33487-2742

© 2006 by Taylor & Francis Group, LLC
CRC Press is an imprint of Taylor & Francis Group

No claim to original U.S. Government works
Printed in the United States of America on acid-free paper
10 9 8 7 6 5 4 3 2 1

International Standard Book Number-10: 0-8493-7278-X (Hardcover)
International Standard Book Number-13: 978-0-8493-7278-0 (Hardcover)
Library of Congress Card Number 2005055971

Library of Congress Cataloging-in-Publication Data

Forsgren, Amy.
    Corrosion control through organic coatings / Amy Forsgren.
        p. cm.
    Includes bibliographical references and index.
    ISBN 0-8493-7278-X (alk. paper)
    1. Protective coatings. 2. Corrosion and anti-corrosives. 3. Organic compounds. I. Title.

TA418.76.F67 2005
620.1'1223--dc22   2005055971

Taylor & Francis Group
is the Academic Division of Informa plc.

Visit the Taylor & Francis Web site at
http://www.taylorandfrancis.com

and the CRC Press Web site at
http://www.crcpress.com

Dedication

---

*To my son Erik and my husband Dr. Per-Ola Forsgren,
without their support and encouragement this book would
not have been possible.*

Preface

This book has been written to fill a gap in the literature of corrosion-protection coatings by offering a bridge between the very brief account of paints conveyed in most corrosion books and the very comprehensive, specialized treatises found in the polymer or electrochemical scientific publications.

I have tried to write this book for the following audiences:

- Maintenance engineers who specify or use anticorrosion paints and need a sound working knowledge of different coating types and some orientation in how to test coatings for corrosion protection
- Buyers or specifiers of coatings, who need to know quickly which tests provide useful knowledge about performance and which do not
- Researchers working with accelerated test methods, who need an in-depth knowledge of aging mechanisms of coatings, in order to develop more accurate tests
- Applicators interested in providing safe working environments for personnel performing surface preparation
- Owners of older steel structures who find themselves faced with removal of lead-based paint (LBP) when carrying out maintenance painting

The subject matter is dictated by the problems all these groups face. LBP dominates parts of the book. Although this coating is on its way out, the problems it has created remain. Replacement pigments of equivalent — even better — quality certainly exist but are not as well known to the general coatings public as we would wish. This is partly due to the chaotic conditions of accelerated testing. Hundreds of test methods exist, with no consensus in the industry about which ones are useful. This confusion has not aided the efforts toward identification and acceptance of the best candidates to replace LBP. And finally, the issues associated with disposal of lead-contaminated blasting debris are expected to become more pressing, not less so, in the future.

However, not all modern maintenance headaches are due to lead. Another problem facing plant engineers and applicators of coatings is silicosis from abrasive blasting with quartz sand. This blasting material is outlawed in many industrialized countries, but sadly, not all. Even in Scandinavia, where worker health is taken very seriously, the ban is not as complete as it should be. And, because we all need the ozone layer, limiting the use of volatile organic compounds where possible is a consideration for today's engineers.

The reader will no doubt notice that, while the book provides plant engineers with a rapid orientation in coating types, abrasives, laboratory techniques, and disposal issues, certain other areas of interest to the same audience are not addressed in this work. Areas such as surface preparation standards, applications methods, and quality control are important and interesting, but in writing a book, it is not possible to include everything. One must draw the line somewhere, and I have chosen to draw it thusly: subjects are not taken up here if they are thoroughly covered in other publications, and the information has already reached a wide audience.

Author

Amy Forsgren received her chemical engineering education at the University of Cincinnati in Ohio in 1986. She then did research in coatings for the paper industry for 3 years, before moving to Detroit, Michigan. There, she spent 6 years in anti-corrosion coatings research at Ford Motor Company, before returning to Sweden in 1996 to lead the protective coatings program at the Swedish Corrosion Institute. In 2001, she joined the telecom equipment industry in Stockholm. Mrs. Forsgren lives in Stockholm with her family.

Acknowledgments

Without the help of many people, this book would not have been possible. I wish in particular to thank my colleague Lars Krantz for generously creating the illustrations. Mats Linder and Bertil Sandberg of the Swedish Corrosion Institute also receive my thanks for supporting the waterborne coatings and lead abatement research programs, as do my colleagues at Semcon AB for taking interest and providing encouragement.

Contents

1 Introduction

This book is not about corrosion; rather it is about paints that prevent corrosion. It was written for those who must protect structural steel from rusting by using anti-corrosion paints. The philosophy of this book is this: if one knows enough about paint, one need not be an expert on rust. In keeping with that spirit, the book endeavors to cover the field of heavy-duty anticorrosion coatings without a single anode or cathode equation explaining the corrosion process. It is enough for us to know that steel will rust if allowed to; we will concentrate on preventing it.

1.1 SCOPE OF THE BOOK

The scope of this book is heavy-duty protective coatings used to protect structural steel, infrastructure components made of steel, and heavy steel process equipment. The areas covered by this book have been chosen to reflect the daily concerns and choices faced by maintenance engineers who use heavy-duty coating, including:

- Composition of anticorrosion coatings
- Waterborne coatings
- Blast-cleaning and other heavy surface pretreatments
- Abrasive blasting and heavy-metal contamination
- Weathering and aging of paint
- Corrosion testing — background and theoretical considerations
- Corrosion testing — practice

1.1.1 TARGET GROUP DESCRIPTION

The target group for this book consists of those who specify, formulate, test, or do research in heavy-duty coatings for such applications as:

- Boxes and girders used under bridges or metal gratings used in the decks of bridges
- Poles for traffic lights and street lighting
- Tanks for chemical storage, potable water, or waste treatment
- Handrails for concrete steps in the fronts of buildings
- Masts for telecommunications antennas
- Power line pylons
- Beams in the roof and walls of food-processing plants
- Grating and framework around processing equipment in paper mills

1

All of these forms of structural steel have at least two things in common:

1. Given a chance, the iron in them will turn to iron oxide.
2. When the steel begins rusting, it cannot be pulled out of service and sent back to a factory for treatment.

During the service life of one of these structures, maintenance painting will have to be done on-site. This imposes certain limitations on the choices the maintenance engineer can make. Coatings that must be applied in a factory cannot be reapplied once the steel is in service. This eliminates organic paints, such as powder coatings or electrodeposition coatings, and several inorganic pretreatments, such as phosphating, hot-dip galvanizing, and chromating. New construction can commonly be protected with these coatings, but they are almost always a one-time-only treatment. When the steel has been in service for a number of years and maintenance coating is being considered, the number of practical techniques is narrowed. This is not to say that the maintenance engineer must face corrosion empty-handed; more good paints are available now than ever before, and the number of feasible pretreatments for cleaning steel in-situ is growing. In addition, coatings users now face such pressures as environmental responsibility in choosing new coatings and disposing of spent abrasives as well as increased awareness of health hazards associated with certain pretreatment methods.

1.1.2 SPECIALTIES OUTSIDE THE SCOPE

Certain anticorrosion coating subspecialties fall outside the scope of this work, including those dealing with automotive, airplane, and marine coatings; powder coatings; and coatings for cathodic protection. These methods are all economically important and scientifically interesting but lie outside of our target group for one or more reasons:

- The way in which the paint is applied can be done only in a factory, so maintenance painting in the field is not possible. (Automotive and powder coatings)
- Aluminium — not steel —is used as the substrate, and the coatings experience temperature extremes and ultraviolent loads that earth-bound structures and their coatings never encounter. (Airplane coatings)
- The circumstances under which marine coatings and coatings with cathodic protection must operate are so different from those experienced by the infrastructure in the target group that different coating and testing technologies are needed. These exist and are already well covered in the technical literature.

1.2 PROTECTION MECHANISMS OF ORGANIC COATINGS

This section presents a brief overview of the various mechanisms by which organic coatings provide corrosion protection to the metal substrate.

Corrosion of a painted metal requires all of the following elements [1]:

- Water
- Oxygen or another reducible species

- A dissolution process at the anode
- A cathode site
- An electrolytic path between the anode and cathode

Any of these items could potentially be rate controlling. A coating that can suppress one or more of the items listed above can therefore limit the amount of corrosion. The main protection mechanisms used by organic coatings are:

- Creating an effective barrier against the corrosion reactants water and oxygen
- Creating a path of extremely high electrical resistance, thus inhibiting anode-cathode reactions
- Passivating the metal surface with soluble pigments
- Providing an alternative anode for the dissolution process

The last two protection mechanisms listed above are discussed extensively in Chapter 2. This section will therefore concentrate on the first two protection mechanisms in the list above.

It must be noted that it is impossible to use all these mechanisms in one coating. For example, pigments whose dissolved ions passivate the metal surface require the presence of water. This rules out their use in a true barrier coating, where water penetration is kept as low as possible.

In addition, the usefulness of each mechanism depends on the service environment. Guruviah studied corrosion of coated panels under various accelerated test methods with and without sodium chloride (salt). Where salt was present, electrolytic resistance of the coatings was the dominant factor in predicting performance. However, in a generally similar method with no sodium chloride, oxygen permeation was the rate-controlling factor for the same coatings [2].

1.2.1 DIFFUSION OF WATER AND OXYGEN

Most coatings, except specialized barrier coatings such as chlorinated rubber, do not protect metal substrates by preventing the diffusion of water. The attractive force for water within most coatings is simply too strong. There seems to be general agreement that the amount of water that can diffuse through organic coatings of reasonable thickness is greater than that needed for the corrosion process [2–8]. Table 1.1 shows the permeation rates of water vapor through several coatings as measured by Thomas [9,10].

The amount of water necessary for corrosion to occur at a rate of 0.07 g Fe/cm^2/year is estimated to be 0.93 g/m^2/day [9,10]. Thus, coatings with the lowest permeability rates might possibly be applied in sufficient thickness such that water does not reach the metal in the amounts needed for corrosion. Other coatings must provide protection through other mechanisms. Similar results have been obtained by other studies [2,11]. However, the role of water permeation through the coating cannot be completely ignored. Haagan and Funke have pointed out that, although water permeability is not normally the rate-controlling step in corrosion, it may be the rate-determining factor in adhesion loss [11].

TABLE 1.1
Water Vapor Permeability

Coating Type	Water Vapor Permeability, g/m²/25µm/day
Chlorinated rubber	20 ± 3
Coat tar epoxy	30 ± 1
Aluminium epoxy mastic	42 ± 6
Red-lead oil-based	214 ± 3
White alkyd	258 ± 6

Sources: Thomas. N.L., *Prog. Org. Coatings*, 19, 101, 1991; Thomas, N.L., *Proc. Symp. Advances in Corrosion Protection by Organic Coatings, Electrochem. Soc.*, 1989, 451.

The amount of oxygen required for a corrosion rate of 0.07 g Fe/cm²/year is estimated to be 575 cc/m²/day. Thomas studied oxygen permeation rates for several types of coatings and found that they have rates far below what is needed to maintain the corrosion reaction, as shown in Table 1.2 [9,10].

These measurements were taken using 1 atmosphere of pure oxygen — that is, nearly five times the amount of oxygen available in air. In Earth's atmosphere, oxygen transport rates may be expected to be lower than this [12]. It should perhaps be noted that these were measurements of oxygen gas permeating through the coating. The amount of oxygen reaching the metal surface will be higher, because water carries dissolved oxygen with it when permeating the coating.

In general, water and oxygen are necessary for the corrosion process; however, their permeation through the coating is not a rate-determining step [13–15].

TABLE 1.2
Oxygen Permeability

Coating Type	Oxygen Permeability, cc/m²/100µm/day
Chlorinated rubber	30 ± 7
Coat tar epoxy	213 ± 38
Aluminium epoxy mastic	110 ± 37
Red-lead oil-based	734 ± 42
White alkyd	595 ± 49

Sources: Thomas. N.L., *Prog. Org. Coatings*, 19, 101, 1991; Thomas, N.L., *Proc. Symp. Advances in Corrosion Protection by Organic Coatings, Electrochem. Soc.*, 1989, 451.

1.2.2 ELECTROLYTIC RESISTANCE

Perhaps the single most important corrosion-protection mechanism of organic coatings is to create a path of extremely high electrical resistance between anodes and cathodes. This electrical resistance reduces the flow of current available for anode-cathode corrosion reactions. In other words, water — but not ions — may readily permeate most coatings. Therefore, the water that reaches a metal substrate is relatively ion-free [12]. Steel corrodes very slowly in pure water, because the ferrous ions and hydroxyl ions form ferrous hydroxide ($Fe(OH)_2$). $Fe(OH)_2$ has low solubility in water (0.0067 g/L at 20° C), precipitates at the site of corrosion, and then inhibits the diffusion necessary to continue corrosion. On the other hand, if chloride or sulphate ions are present, they react with steel to form ferrous chloride and sulphate complexes. These are soluble and can diffuse away from the site of corrosion. After diffusing away, they can be oxidized, hydrolyzed, and precipitated as rust some distance away from the corrosion site. The stimulating Cl^- or SO_4^{2-} anion is liberated and can re-enter the corrosion cycle until it becomes physically locked up in insoluble corrosion products [16-21]. This mechanism of blocking ions has several names, including electrolytic resistance, resistance inhibition, and ionic resistance. The terms *electrolytic resistance* and *ionic resistance* are used more-or-less interchangeably, because Kittleberger and Elm showed a linear relationship between the diffusion of ions and the reciprocal of the film resistance [22].

Overall, the electrolytic resistance of an immersed coating can be said to depend on at least two factors: the activity of the water in which the coating is immersed and the nature of the counter ion inside the polymer [1]. Bacon and colleagues have performed extensive work establishing the correlation between electrolytic (ionic) resistance of the coating and its ability to protect the steel substrate from corrosion. In a study involving more than 300 coating systems, they observed good corrosion protection in coatings that could maintain a resistance of 108 Ω/cm^2 over an exposure period of several months; they did not observe the same results in coatings whose resistance fell below this [23].

Mayne deduced the importance of electrolytic resistance as a protection mechanism from the high rates of water and oxygen transport through coatings. Specifically, Mayne and coworkers [7, 24-27] found that the resistance of immersed coatings could change over time. From their studies, they concluded that at least two processes control the ionic resistance of immersed coatings:

- A fast change, which takes place within minutes of immersion
- A slow change, which takes weeks or months [26]

The fast change is related to the amount of water in the film. Its controlling factor is osmotic pressure. The slow change is controlled by the concentration of electrolytes in the immersion solution. An exchange of cations in the electrolyte for hydrogen ions in the coating may lie behind this steady fall, over months, in the coating resistance. This theory has received some support from the work of Khullar and Ulfvarson, who found an inverse relationship between the ion exchange capacity and the corrosion protection efficiency of paint films [13, 28]. The structural changes brought about by this ion exchange might slowly destroy the protective properties of the film [29].

Many workers in the field of water transport have concentrated on the physical properties of film, such as capillary structure, or composition of the electrolytes. The work of Kumins and London has shown that the chemical composition of the polymer is equally important. In particular, the concentration of fixed anions in the polymer film is critical. They found that if the concentration of salt in the electrolyte was below the film's fixed-anion concentration, the passage of anions through the film was very restricted. If the electrolyte's concentration was above the polymer's fixed-anion concentration, anions could permeate much more freely through the film [30].

Further information regarding the mechanisms of ion transport through the coating film can be found in reviews by Koehler, Walter, and Greenfield and Scantlebury [1, 29, 31].

1.2.3 ADHESION

When a metal substrate has corroded, the paint no longer adheres to it. Accordingly, corrosion workers commonly place heavy emphasis on the importance of adhesion of the organic coating to the metal substrate, and a great deal of energy has gone into developing test methods for quantifying this adhesion.

1.2.3.1 What Adhesion Accomplishes

Very strong adhesion can help suppress corrosion by resisting the development of corrosion products, hydrogen evolution, or water build-up under the coating [32-35]. In addition, by bonding to as many available active sites on the metal surface as possible, the coating acts as an electrical insulator, thereby suppressing the formation of anode-cathode microcells among inhomogeneities in the surface of the metal.

The role of adhesion is to create the necessary conditions so that corrosion-protection mechanisms can work. A coating cannot passivate the metal surface, create a path of extremely high electrical resistance at the metal surface, or prevent water or oxygen from reaching the metal surface unless it is in intimate contact — at the atomic level — with the surface. The more chemical bonds between the surface and coating, the closer the contact and the stronger the adhesion. An irreverent view could be that the higher the number of sites on the metal that are taken up in bonding with the coating, the lower the number of sites remaining available for electrochemical mischief. Or as Koehler expressed it:

> The position taken here is that from a corrosion standpoint, the degree of adhesion is in itself not important. It is only important that some degree of adhesion to the metal substrate be maintained. Naturally, if some external agency causes detachment of the organic coating and there is a concurrent break in the organic coating, the coating will no longer serve its function over the affected area. Typically, however, the detachment occurring is the result of the corrosion processes and is not quantitatively related to adhesion [1].

In summary, good adhesion of the coating to the substrate could be described as a "necessary but not sufficient" condition for good corrosion protection. For all of the protection mechanisms described in the previous sections, good adhesion of the coating to the metal is a necessary condition. However, good adhesion alone is not enough; adhesion tests in isolation cannot predict the ability of a coating to control corrosion [36].

1.2.3.2 Wet Adhesion

A coating can be saturated with water, but if it adheres tightly to the metal, it can still prevent sufficient amounts of electrolytes from collecting at the metal surface for the initiation of corrosion. How well the coating clings to the substrate when it is saturated is known as *wet adhesion*. Adhesion under dry conditions is probably overrated; wet adhesion, on the other hand, is crucial to corrosion protection.

Commonly, coatings with good dry adhesion have poor wet adhesion [37-41]. The same polar groups on the binder molecules that create good dry adhesion can wreak mischief by decreasing water resistance at the coating-metal interface — that is, they decrease wet adhesion [42]. Another important difference is that, once lost, dry adhesion cannot be recovered. Loss of adhesion in wet conditions, on the other hand, can be reversible, although the original dry adhesion strength will probably not be obtained [16, 43].

Perhaps it should be noted that wet adhesion is a coating property and not a failure mechanism. Permanent adhesion loss due to humid or wet circumstances also exists and is called *water disbondment.*

Relatively little research has been done on wet adhesion phenomena. Leidheiser has identified some important questions in this area [43]:

1. How can wet adhesion be quantitatively measured while the coating is wet?
2. What is the governing principle by which water collects at the organic coating-metal interface?
3. What is the thickness of the water layer at the interface, and what determines this thickness?

Two additional questions could be added to this list:

4. What makes adhesion loss under wet circumstances irreversible? Is there a relationship between the coating property, wet adhesion, the failure mechanism, and water disbondment?
5. Why does the reduction of adhesion on exposure to water not lead to complete delamination? What causes residual adhesion in wet circumstances?

As a possible answer to the last question above, Funke has suggested that dry adhesion is due to a mixture of bond types. Polar bonding, which is somewhat sensitive to water molecules, could account for reduced adhesion in wet circumstances, whereas chemical bonds or mechanical locking may account for residual adhesion [16]. Further research on wet adhesion could answer some of the aforementioned questions and increase understanding of this complex mechanism.

1.2.3.3 Important Aspects of Adhesion

Two aspects of adhesion are important: the initial strength of the coating-substrate bond and what happens to this bond as the coating ages.

A great deal of work has been done to develop better methods for measuring the initial strength of the coating-substrate bond. Unfortunately, the emphasis on measuring initial adhesion may miss the point completely. It is certainly true that good adhesion between the metal and the coating is necessary for preventing corrosion under the coating. However, it is possible to pay too much attention to measuring the difference between very good initial adhesion and excellent initial adhesion, completely missing the question of whether or not that adhesion is maintained. In other words, as long as the coating has good initial adhesion, then it does not matter whether that adhesion is simply very good or great. What matters is what happens to the adhesion over time. This aspect is much more crucial to coating success or failure than is the exact value of the initial adhesion.

Adhesion tests on aged coatings are useful not only to ascertain if the coatings still adhere to the metal but also to yield information about the mechanisms of coating failure. This area deserves greater attention, because studying changes in the failure loci in adhesion tests before and after weathering can yield a great deal of information about coating deterioration.

1.2.4 PASSIVATING WITH PIGMENTS

Anticorrosion pigments in a coating dissolve in the presence of water. Their dissociated ions migrate to the coating-metal interface and passivate it by supporting the formation of thin layers of insoluble corrosion products, which inhibit further corrosion [44-46]. For more information about anticorrosion pigments, see Chapter 2.

1.2.5 ALTERNATIVE ANODES (CATHODIC PROTECTION)

Some very effective anticorrosion coatings allow the conditions necessary for corrosion to occur — for example, water, oxygen, and ions may be present; the coating does not offer much electrical resistance; or soluble pigments have not passivated the metal surface. These coatings do not protect by suppressing the corrosion process; rather, they provide another metal that will corrode instead of the substrate. This mechanism is referred to as *cathodic protection*. In protective coatings, the most important example of cathodic protection is zinc-rich paints, whose zinc pigment acts as a sacrificial anode, corroding preferentially to the steel substrate. For more information on zinc-rich coatings, see Chapter 2.

REFERENCES

1. Koehler, E.L., *Corrosion under organic coatings,* Proc., U.R.. Evans International Conference on Localized Corrosion, NACE, Houston, 1971, 117.
2. Guruviah, S., *JOCCA,* 53, 669, 1970.
3. Mayne, J.E.O., *JOCCA,* 32, 481, 1949.
4. Thomas, A.M and Gent, W.L., *Proc. Phys. Soc.,* 57, 324, 1945.
5. Anderson, A.P. and Wright, K.A., *Industr. Engng. Chem.,* 33, 991, 1941.
6. Edwards, J.D. and Wray, R.I., *Industr. Engng. Chem.,* 28, 549, 1936
7. Maitland, C.C. and Mayne, J.E.O., *Off. Dig.,* 34, 972, 1962.
8. McSweeney, E.E., *Off. Dig.,* 37, 626, 1965.

9. Thomas. N.L., Prog. *Org. Coat.*, 19, 101, 1991.
10. Thomas, N.L., Proc. Symp. Advances in Corrosion Protection by Organic Coatings, *Electrochem. Soc.*, 1989, 451.
11. Haagen, H. and Funke, W., *JOCCA*, 58, 359. 1975.
12. Wheat, N., *Prot. Coat. Eur.*, 3, 24, 1998.
13. Khullar, M.L. and Ulfvarson, U., Proc., IXth FATIPEC Congress, Fédération d'Associations de Techniciens des Industries des Peintures, Vernis, Emaux et Encres d'Imprimerie de l'Europe Continentale (FATIPEC), Paris, 1968, 165.
14. Bacon, C. et al., *Ind. Eng. Chem.*, 161, 40, 1948.
15. Cherry, B.W., *Australag. Corr. and Eng.*, 10, 18, 1974.
16. Funke, W., in *Surface Coatings – 2*, Wilson, A.D., Nicholson, J.W. and Prosser, H.J., Eds., Elsevier Applied Science, London, 1988, 107.
17 Kaesche, H., *Werkstoffe u. Korrosion*, 15, 379, 1964.
18. Knotkowa-Cermakova, A. and Vlekova, J., *Werkstoffe u. Korrosion*, 21, 16, 1970.
19. Schikorr, G., *Werkstoffe u. Korrosion*, 15, 457, 1964.
20. Dunkan, J.R., *Werkstoffe u. Korrosion*, 25, 420, 1974.
21. Barton, K. and Beranek, E., *Werkstoffe u. Korrosion*, 10, 377, 1959.
22. Kittleberger, W.W. and Elm, A.C. *Ind. Eng. Chem.*, 44, 326, 1952.
23. Bacon, C.R., Smith, J.J. and Rugg, F.M., *Ind. Eng. Chem.*, 40, 161, 1948.
24. Cherry, B.W. and Mayne, J.E.O., Proc., First International Congress on Metallic Corrosion, Butterworths, London, 1961.
25. Mayne, J.E.O., *Trans. Inst. Met. Finish.*, 41, 121, 1964.
26. Cherry, B.W. and Mayne, J.E.O. *Off. Dig.*, 37, 13, 1965.
27. Mayne, J.E.O., *JOCCA*, 40, 183, 1957.
28. Ulfvarson, U. and Khullar, M., *JOCCA*, 54, 604, 1971.
29. Walter, G.W., *Corros. Science*, 26, 27, 1986.
30. Kumins, C.A. and London, A., J. *Polymer Science*, 46, 395, 1960.
31. Greenfield, D. and Scantlebury, D., *J. Corros. Science and Eng.*, 3, Paper 5, 2000.
32. Patrick, R.L. and Millar, R.L. in *Handbook of Adhesives*, Skeist, I., Ed. Reinhold Publishing Corp., New York, 1962, 602.
33. Kittleberger, W.W. and Elm, A.C., *Ind. Eng. Chem.*, 38, 695, 1946.
34. Evans, U.R. *Corrosion and Oxidation of Metals*, St. Martin's Press, New York. 1960.
35. Gowers, K.R. and Scantlebury, J.D. *JOCCA*, 4, 114, 1988.
36. Troyk, P.R., Watson, M.J. and Poyezdala, J.J., in ACS Symposium Series 322: Polymeric Materials for Corrosion Control, Dickie, R.A. and Floyd, F.L, Eds., American Chemical Society, Washington DC, 1986, 299.
37. Bullett, T.R., *JOCCA*, 46, 441, 1963.
38. Walker, P., *Off. Dig.*, 37, 1561, 1965.
39. Walker, P., *Paint Technol.*, 31, 22, 1967.
40. Walker, P., *Paint Technol.*, 31, 15, 1967.
41. Funke, W., *J. Coat. Technol.*, 55, 31, 1983.
42. Funke, W., in ACS Symposium Series 322: Polymeric Materials for Corrosion Control, Dickie, R.A. and Floyd, F.L, Eds., American Chemical Society, Washington DC, 1986, 222.
43. Leidheiser, H., in ACS Symposium Series 322: Polymeric Materials for Corrosion Control, Dickie, R.A. and Floyd, F.L, Eds., American Chemical Society, Washington DC, 1986, 124.
44. Mayne, J.E.O. and Ramshaw, E.H. J. *Appl. Chem.*, 13, 553, 1969.
45. Leidheiser, H., J. *Coat. Technol.*, 53, 29, 1981.
46. Mayne, J.E.O., in *Pigment Handbook*, Vol. III, Patton, T.C., Ed. Wiley Interscience Publ., 1973, 457.

2 Composition of the Anticorrosion Coating

2.1 COATING COMPOSITION DESIGN

Generally, the formulation of a coating may be said to consist of the binder, pigment, additives, and carrier. The binder and the pigment are the most important elements; they may be said to perform the corrosion-protection work in the cured paint.

With very few exceptions (e.g., inorganic zinc-rich primers [ZRPs]), binders are organic polymers. A combination of polymers is frequently used, even if the coating belongs to one generic class. An acrylic paint, for example, may purposely use several acrylics derived from different monomers or from similar monomers with varying molecular weights and functional groups of the final polymer. Polymer blends capitalize on each polymer's special characteristics; for example, a methacrylate-based acrylic with its excellent hardness and strength should be blended with a softer polyacrylate to give some flexibility to the cured paint.

Pigments are added for corrosion protection, for color, and as filler. Anticorrosion pigments are chemically active in the cured coating, whereas pigments in barrier coatings must be inert. Filler pigments must be inert at all times, of course, and the coloring of a coating should stay constant throughout its service life.

Additives may alter certain characteristics of the binder, pigment, or carrier to improve processing and compatibility of the raw materials or application and cure of the coating.

The carrier is the vehicle in the uncured paint that carries the binder, the pigments, and the additives. It exists only in the uncured state. Carriers are liquids in the case of solvent-borne and waterborne coatings, and gases in the case of powder coatings.

2.2 BINDER TYPES

The binder of a cured coating is analogous to the skeleton and skin of the human body. In the manner of a skeleton, the binder provides the physical structure to support and contain the pigments and additives. It binds itself to these components and to the metal surface, hence its name. It also acts somewhat as a skin: the amounts of oxygen, ions, water, and ultraviolet (UV) radiation that can penetrate into the cured coating layer depend to some extent on which polymer is used. This is because the cured coating is a very thin polymer-rich or pure polymer layer over a heterogeneous mix of pigment particles and binder. The thin topmost layer — sometimes known as the *healed layer* of the coating — covers gaps between pigment particles and cured binder, through which water finds its easiest route to the metal surface. It can also cover pores in the bulk of the coating, blocking this means of water transport. Because this healed surface is very thin, however, its ability to entirely prevent water uptake is greatly limited. Generally, it succeeds much better at limiting transport of oxygen. The ability to absorb, rather than transmit, UV radiation is polymer-dependent; acrylics, for example, are for most purposes impervious to UV-light, whereas epoxies are extremely sensitive to it.

The binders used in anticorrosion paints are almost exclusively organic polymers. The only commercially significant exceptions are the silicon-based binder in inorganic ZRPs sil oxanes, and high-temperature silicone coatings. Many of the coating's physical and mechanical properties — including flexibility, hardness, chemical resistances, UV-vulnerability, and water and oxygen transport — are determined wholly or in part by the particular polymer or blend of polymers used.

Combinations of monomers and polymers are commonly used, even if the coating belongs to one generic polymer class. Literally hundreds of acrylics are commercially available, all chemically unique; they differ in molecular weights, functional groups, starting monomers, and other characteristics. A paint formulator may purposely blend several acrylics to take advantage of the characteristics of each; thus a methacrylate-based acrylic with its excellent hardness and strength might be blended with one of the softer polyacrylates to impart flexibility to the cured paint.

Hybrids, or combinations of different polymer families, are also used. Examples of hybrids include acrylic-alkyd hybrid waterborne paints and the epoxy-modified alkyds known as epoxy ester paints.

2.2.1 Epoxies

Because of their superior strength, chemical resistance, and adhesion to substrates, epoxies are the most important class of anticorrosive paint. In general, epoxies have the following features:

- Very strong mechanical properties
- Very good adhesion to metal substrates
- Excellent chemical, acid, and water resistance
- Better alkali resistance than most other types of polymers
- Susceptibility to UV degradation

2.2.1.1 Chemistry

The term *epoxy* refers to thermosetting polymers produced by reaction of an epoxide group (also known as the glycidyl, epoxy, or oxirane group; see Figure 2.1). The ring structure of the epoxide group provides a site for crosslinking with proton donors, usually amines or polyamides [1].

FIGURE 2.1 Epoxide or oxirane group.

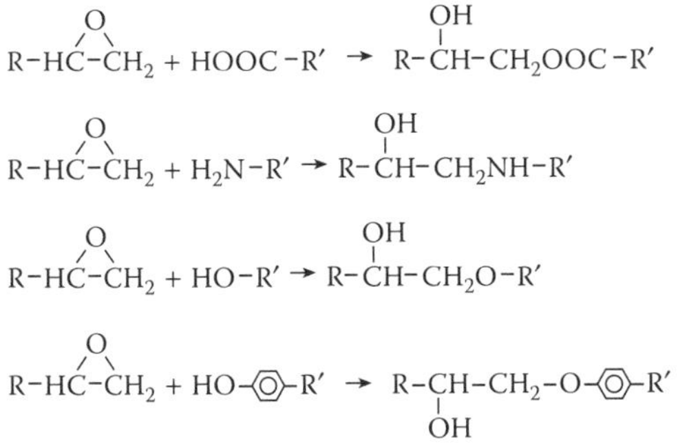

$$R-\overset{O}{\overset{/\backslash}{HC}-CH_2} + HOOC-R' \;\rightarrow\; R-\overset{OH}{\overset{|}{CH}}-CH_2OOC-R'$$

$$R-\overset{O}{\overset{/\backslash}{HC}-CH_2} + H_2N-R' \;\rightarrow\; R-\overset{OH}{\overset{|}{CH}}-CH_2NH-R'$$

$$R-\overset{O}{\overset{/\backslash}{HC}-CH_2} + HO-R' \;\rightarrow\; R-\overset{OH}{\overset{|}{CH}}-CH_2O-R'$$

$$R-\overset{O}{\overset{/\backslash}{HC}-CH_2} + HO-\underset{}{\bigcirc}-R' \;\rightarrow\; R-\underset{\underset{OH}{|}}{CH}-CH_2-O-\bigcirc-R'$$

FIGURE 2.2 Typical reactions of the epoxide (oxirane) group to form epoxies.

Epoxies have a wide variety of forms, depending on whether the epoxy resin (which contains the epoxide group) reacts with a carboxyl, hydroxyl, phenol, or amine curing agent. Some of the typical reactions and resulting polymers are shown in Figure 2.2. The most commonly used epoxy resins are [2]:

- Diglycidyl ethers of bisphenol A (DGEBA or Bis A epoxies)
- Diglycidyl ethers of bisphenol F (DGEBF or Bis F epoxies) — used for low-molecular-weight epoxy coatings
- Epoxy phenol or cresol novolac multifunctional resins

Curing agents include [2]:

- Aliphatic polyamines
- Polyamine adducts
- Ketimines
- Polyamides/amidoamines
- Aromatic amines
- Cycloaliphatic amines
- Polyisocyanates

2.2.1.2 Ultraviolet Degradation

Epoxies are known for their susceptibility to UV degradation. The UV rays of the sun contain enough energy to break certain bonds in the polymeric structure of a cured binder. As more and more bond breakage occurs in the top surface of the cured binder layer, the polymeric backbone begins to break down. Because the topmost surface or "healed layer" of the cured coating contains only binder, the initial result of the UV degradation is simply loss of gloss. However, as the degradation works downward through the coating layer, binder breakdown begins to free pigment particles. A fine powder consisting of pigment and fragments of binder continually forms on the surface of the coating. The powder is reminiscent of chalk dust, hence the name "chalking" for this breakdown process.

Chalking also occurs to some extent with several other types of polymers. It does not directly affect corrosion protection but is a concern because it eventually results in a thinner coating. The problem is easily overcome with epoxies, however, by covering the epoxy layer with a coating that contains a UV-resistant binder. Polyurethanes are frequently used for this purpose because they are similar in chemical structure to epoxies but are not susceptible to UV breakdown.

2.2.1.3 Variety of Epoxy Paints

The resins used in the epoxy reactions described in section 2.2.1.1 are available in a wide range of molecular weights. In general, as molecular weight increases, flexibility, adhesion, substrate wetting, pot life, viscosity, and toughness increase. Increased molecular weight also corresponds to decreased crosslink density, solvent resistance, and chemical resistance [2]. Resins of differing molecular weights are usually blended to provide the balance of properties needed for a particular type of coating.

The number of epoxide reactions possible is practically infinite and has resulted in a huge variety of epoxy polymers. Paint formulators have taken advantage of this variability to provide epoxy paints with a wide range of physical, chemical, and mechanical characteristics. The term "epoxy" encompasses an extremely wide range of coatings, from very-low-viscosity epoxy sealers (for penetration of crevices) to exceptionally thick epoxy mastic coatings.

2.2.1.3.1 Mastics

Mastics are high-solids, high-build epoxy coatings designed for situations in which surface preparation is less than ideal. They are sometimes referred to as "surface tolerant" because they do not require the substrate to be cleaned by abrasive blasting to Sa2 1/2. Mastics can tolerate a lack of surface profile (for anchoring) and a certain amount of contamination (e.g., by oil) that would cause other types of paints to quickly fail.

Formulation is challenging, because the demands placed on this class can be contradictory. Because they are used on smoother and less clean surfaces, mastics must have good wetting characteristics. At the same time, viscosity must be very high to prevent sagging of the very thick wet film on vertical surfaces. Meeting both of these requirements presents a challenge to the paint chemist.

Epoxy mastics with aluminium flake pigments have very low moisture permeations and are popular both as spot primers or full coats. They can be formulated with weak solvents and thus can be used over old paint. The lack of aggressive solvents in mastics means that old paints will not be destroyed by epoxy mastics. This characteristic is needed for spot primers, which overlap old, intact paint at the edge of the spot to be coated. Mastics pigmented with aluminium flake are also used as full-coat primers.

Because of their very high dry film thickness, build-up of internal stress in the coating during cure is often an important consideration in using mastic coatings.

2.2.1.3.2 Solvent-Free Epoxies

Another type of commonly used epoxy paint is the solvent-free, or 100% solid, epoxies. Despite their name, these epoxies are not completely solvent-free. The levels of organic solvents are very low, typically below 5%, which allows very high film builds and greatly reduces concerns about volatile organic compounds (VOCs).

An interesting note about these coatings is that many of them generate significant amounts of heat upon mixing. The cross-linking is exothermic, and little solvent is present to take up the heat in vaporization [2].

2.2.1.3.3 Glass Flake Epoxies

Glass flake epoxy coatings are used to protect steel in extremely aggressive environments. When these coatings were first introduced, they were primarily used in offshore applications. In recent years, however, they have been gaining acceptance in mainstream infrastructure as well. Glass flake pigments are large and very thin, which allows them to form many dense layers with a large degree of overlap between glass particles. This layering creates a highly effective barrier against moisture and chemical penetration because the pathway around and between the glass flakes is extremely long. The glass pigment can also confer increased impact and abrasion resistance and may aid in relieving internal stress in the cured coating.

2.2.1.3.4 Coal Tar Epoxies

Coal tar, or pitch, is the black organic resin left over from the distillation of coal. It is nearly waterproof and has been added to epoxy amine and polymide paints to obtain coatings with very low water permeability. It should be noted that coal tar products contain polynuclear aromatic compounds, which are suspected to be carcinogenic. The use of coal tar coatings is therefore restricted or banned in some countries.

2.2.2 ACRYLICS

Acrylics is a term used to describe a large and varied family of polymers. General characteristics of this group include:

- Outstanding UV stability
- Good mechanical properties, particularly toughness [3]

Their exceptional UV resistance makes acrylics particularly suitable for applications in which retention of clarity and color are important.

Acrylic polymers can be used in both waterborne and solvent-borne coating formulations. For anticorrosion paints, the term *acrylic* usually refers to waterborne or latex formulations.

2.2.2.1 Chemistry

Acrylics are formed by radical polymerization. In this chain of reactions, an initiator — typically a compound with an azo link (—N=N—) or a peroxy link (— 0–0—) — breaks down at the central bond, creating two free radicals. These free radicals combine with a monomer, creating a larger free-radical molecule, which continues to grow as it combines with monomers, until it either:

- Combines with another free radical (effectively canceling each other)
- Reacts with another free radical: briefly meeting, transferring electrons and splitting unevenly, so that one molecule has an extra hydrogen atom and one is lacking a hydrogen atom (a process known as disproportionation)

TABLE 2.1
Main Reactions Occurring in Free Radical Chain Addition Polymerization

	Radical Polymerization I = Initiator; M = Monomer
Initiator breakdown	$I{:}I \rightarrow I + I$
Initiation and propagation	$I + M_n \rightarrow I(M)_n$
Termination by combination	$I(M)_n + (M)_m I \rightarrow I(M)_{m+n} I$
Termination by disproportionation	$I(M)_n + (M)_n I \rightarrow I(M)_{n-1+n}(M{-}H) + I(M)_{m-1}(M{+}H)$

Data from: Bentley, J., Organic film formers, in *Paint and Surface Coatings Theory and Practice*, Lambourne, R., Ed., Ellis Horwood Limited, Chichester, 1987.

- Transfers the free radical to another polymer, a solvent, or a chain transfer agent, such as a low-molecular-weight mercaptan to control molecular weight

This process, excluding transfer, is depicted in Table 2.1 [4].
Some typical initiators used are listed here and shown in Figure 2.3.

- Azo di isobutyronitrile (AZDN)
- Di benzoyl peroxide
- *T*-butyl perbenzoate
- Di *t*-butyl peroxide

Typical unsaturated monomers include:

- Methacrylic acid
- Methyl methacrylate
- Butyl methacrylate
- Ethyl acrylate
- 2-Ethyl hexyl acrylate

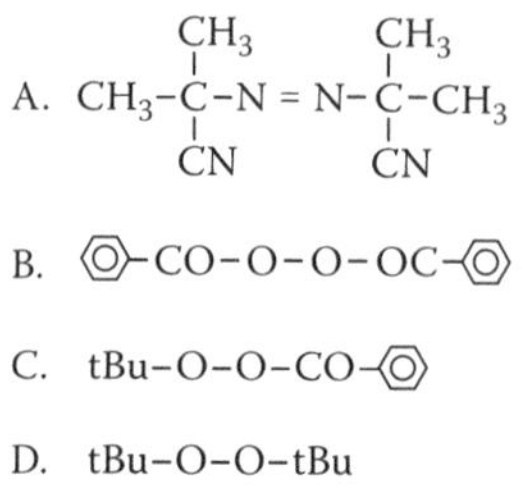

$$
\begin{array}{c}
\qquad\quad CH_3 \qquad\ CH_3 \\
\qquad\qquad | \qquad\qquad\ | \\
A.\ \ CH_3{-}C{-}N = N{-}C{-}CH_3 \\
\qquad\qquad | \qquad\qquad\ | \\
\qquad\quad CN \qquad\quad CN
\end{array}
$$

B. $\bigcirc\!\!-CO{-}O{-}O{-}OC\!\!-\!\bigcirc$

C. $tBu{-}O{-}O{-}CO\!\!-\!\bigcirc$

D. $tBu{-}O{-}O{-}tBu$

FIGURE 2.3 Typical initiators in radical polymerization: A = AZDN; B = Di benzoyl peroxide; C = *T*-butyl perbenzoate; D = Di *t*-butyl peroxide.

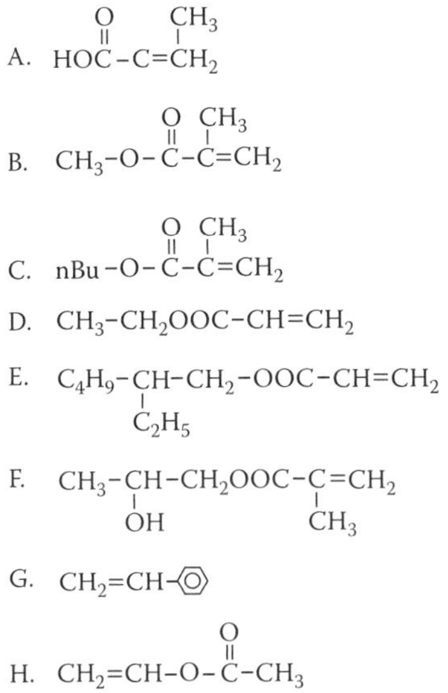

$$
\begin{array}{ll}
\text{A.} & \underset{\displaystyle \text{HO}\overset{\displaystyle O}{\overset{\|}{C}}-\overset{\displaystyle CH_3}{\underset{}{C}}=CH_2}{}
\end{array}
$$

A.   HOC-C=CH₂ (with O above as ‖ on C and CH₃ above the central C)

B.   CH₃-O-C-C=CH₂ (with O as ‖ on first C and CH₃ above second C)

C.   nBu-O-C-C=CH₂ (with O as ‖ and CH₃ above)

D.   CH₃-CH₂OOC-CH=CH₂

E.   C₄H₉-CH-CH₂-OOC-CH=CH₂
            |
           C₂H₅

F.   CH₃-CH-CH₂OOC-C=CH₂
            | |
           OH   CH₃

G.   CH₂=CH-⬡

H.   CH₂=CH-O-C-CH₃ (with O as ‖ on C)

FIGURE 2.4 Typical unsaturated monomers: A = Methacrylic acid; B = Methyl methacrylate; C = Butyl methacrylate; D = Ethyl acrylate; E = 2-Ethyl hexyl acrylate; F = 2-Hydroxy propyl methacrylate; G = Styrene; H = Vinyl acetate.

- 2-Hydroxy propyl methacrylate
- Styrene
- Vinyl acetate (see also Figure 2.4)

2.2.2.2 Saponification

Acrylics can be somewhat sensitive to alkali environments — such as those which can be created by zinc surfaces [5]. This sensitivity is nowhere near as severe as those of alkyds and is easily avoided by proper choice of copolymers.

Acrylics can be divided into two groups, acrylates and methacrylates, depending on the original monomer from which the polymer was built. As shown in Figure 2.5, the difference lies in a methyl group attached to the backbone of the polymer molecule of a methacrylate in place of the hydrogen atom found in the acrylate.

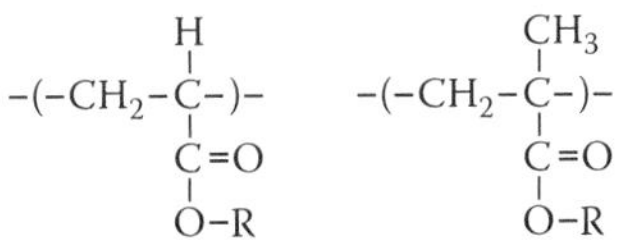

$$
\begin{array}{cc}
\text{H} & \text{CH}_3 \\
| & | \\
-(-CH_2-C-)- & -(-CH_2-C-)- \\
| & | \\
C=O & C=O \\
| & | \\
O-R & O-R
\end{array}
$$

FIGURE 2.5 Depiction of an acrylate (left) and a methacrylate (right) polymer molecule.

Poly(methyl methacrylate) is quite resistant to alkaline saponification; the problem lies with the polyacrylates [6]. However, acrylic emulsion polymers cannot be composed solely of methyl methacrylate because the resulting polymer would have a minimum film formation temperature of over 100°C. Forming a film at room temperature with methyl methacrylate would require unacceptably high amounts of external plasticizers or coalescing solvents. For paint formulations, acrylic emulsion polymers must be copolymerized with acrylate monomers.

Acrylics can be successfully formulated for coating zinc or other potentially alkali surfaces, if careful attention is given to the types of monomer used for copolymerization.

2.2.2.3 Copolymers

Most acrylic coatings are copolymers, in which two or more acrylic polymers are blended to make the binder. This practice combines the advantages of each polymer. Poly(methyl methacrylate), for example, is resistant to saponification, or alkali breakdown. This makes it a highly desirable polymer for coating zinc substrates or any surfaces where alkali conditions may arise. Certain other properties of methyl methacrylate, however, require some modification from a copolymer in order to form a satisfactory paint. For example, the elongation of pure methyl methacrylate is undesirably low for both solvent-borne and waterborne coatings (see Table 2.2) [7]. A "softer" acrylate copolymer is therefore used to impart to the binder the necessary ability to flex and bend. Copolymers of acrylates and methacrylates can give the binder the desired balance between hardness and flexibility. Among other properties, acrylates give the coating improved cold crack resistance and adhesion to the substrate, whereas methacrylates contribute toughness and alkali resistance [3,4,6]. In waterborne formulations, methyl methacrylate emulsion polymers alone could not form films at room temperature without high amounts of plasticizers, coalescing solvents, or both.

Copolymerization is also used to improve solvent and water release in the wet stage, and resistance to solvents and water absorption in the cured coating. Styrene is used for hardness and water resistance, and acrylonitrile imparts solvent resistance [3].

TABLE 2.2
Mechanical Properties of Methyl Methacrylate and Polyacrylates

	Methyl methacrylate	Polyacrylates
Tensile strength (psi)	9000	3-1000
Elongation at break	4%	750-2000%

Modified from: Brendley, W.H., *Paint and Varnish Production,* 63, 19, 1973.

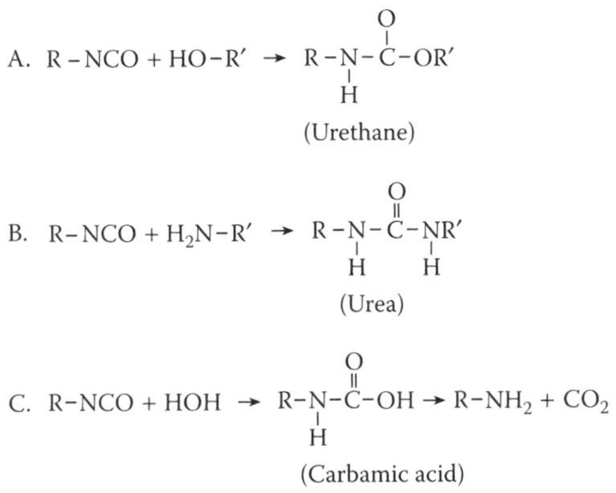

$$
\begin{array}{c}
\overset{O}{\underset{|}{}}\\
\text{A. R-NCO + HO-R'} \rightarrow \text{R-N-C-OR'}\\
\underset{H}{|}
\end{array}
$$

(Urethane)

$$
\begin{array}{c}
\overset{O}{\underset{\|}{}}\\
\text{B. R-NCO + H}_2\text{N-R'} \rightarrow \text{R-N-C-NR'}\\
\underset{H}{|}\underset{H}{|}
\end{array}
$$

(Urea)

$$
\begin{array}{c}
\overset{O}{\underset{\|}{}}\\
\text{C. R-NCO + HOH} \rightarrow \text{R-N-C-OH} \rightarrow \text{R-NH}_2 + \text{CO}_2\\
\underset{H}{|}
\end{array}
$$

(Carbamic acid)

FIGURE 2.6 Some typical isocyanate reactions. A-hydroxyl reaction; B-amino reaction; C-moisture core reaction.

2.2.3 POLYURETHANES

Polyurethanes as a class have the following characteristics:

*   Excellent water resistance [1]
*   Good resistance to acids and solvents
*   Better alkali resistance than most other polymers
*   Good abrasion resistance and, in general, good mechanical properties

They are formed by isocyanate (R–N=C=O) reactions, typically with hydroxyl groups, amines, or water. Some typical reactions are shown in Figure 2.6. Polyurethanes are classified into two types, depending on their curing mechanisms: moisture-cure urethanes and chemical-cure urethanes [1]. These are described in more detail in subsequent sections. Both moisture-cure and chemical-cure polyurethanes can be made from either aliphatic or aromatic isocyanates.

Aromatic polyurethanes are made from isocyanates that contain unsaturated carbon rings, for example, toluene diisocyanate. Aromatic polyurethanes cure faster due to inherently higher chemical reactivity of the polyisocyanates [8], have more chemical and solvent resistance, and are less expensive than aliphatics but are more susceptible to UV radiation [1,9,10]. They are mostly used, therefore, as primers or intermediate coats in conjunction with nonaromatic topcoats that provide UV protection. The UV susceptibility of aromatic polyurethane primers means that the time that elapses between applying coats is very important. The manufacturer's recommendations for maximum recoat time should be carefully followed.

Aliphatic polyurethanes are made from isocyanates that do not contain unsaturated carbon rings. They may have linear or cyclic structures; in cyclic structures, the ring is saturated [11]. The UV resistance of aliphatic polyurethanes is higher than that of aromatic polyurethanes, which results in better weathering characteristics, such as gloss and color retention. For outdoor applications in which good weatherability is necessary, aliphatic topcoats are preferable [1,9]. In aromatic-aliphatic blends, even small amounts of an aromatic component can significantly affect gloss retention [12].

2.2.3.1 Moisture-Cure Urethanes

Moisture-cure urethanes are one-component coatings. The resin has at least two isocyanate groups (–N=C=O) attached to the polymer. These functional groups react with anything containing reactive hydrogen, including water, alcohols, amines, ureas, and other polyurethanes. In moisture-cure urethane coatings, some of the isocyanate reacts with water in the air to form carbamic acid, which is unstable. This acid decomposes to an amine which, in turn, reacts with other isocyanates to form a urea. The urea can continue reacting with any available isocyanates, forming a biuret structure, until all the reactive groups have been consumed [9,11]. Because each molecule contains at least two –N=C=O groups, the result is a crosslinked film.

Because of their curing mechanism, moisture-cure urethanes are tolerant of damp surfaces. Too much moisture on the substrate surface is, of course, detrimental, because isocyanate reacts more easily with water rather than with reactive hydrogen on the substrate surface, leading to adhesion problems. Another factor that limits how much water can be tolerated on the substrate surface is carbon dioxide (CO_2). CO_2 is a product of isocyanate's reaction with water. Too rapid CO_2 production can lead to bubbling, pinholes, or voids in the coating [9].

Pigmenting moisture-cure polyurethanes is not easy because, like all additives, pigments must be free from moisture [9]. The color range is therefore somewhat limited compared with the color range of other types of coatings.

2.2.3.2 Chemical-Cure Urethanes

Chemical-cure urethanes are two-component coatings, with a limited pot life after mixing. The reactants in chemical-cure urethanes are:

1. A material containing an isocyanate group (–N=C=O)
2. A substance bearing free or latent active hydrogen-containing groups (i.e., hydroxyl or amino groups) [8]

The first reactant acts as the curing agent. Five major monomeric diisocyanates are commercially available [10]:

- Toluene diisocyanate (TDI)
- Methylene diphenyl diisocyanate (MDI)
- Hexamethylene diisocyanate (HDI)
- Isophorone diisocyanate (IPDI)
- Hydrogenated MDI ($H_{12}MDI$)

The second reactant is usually a hydroxyl-group-containing oligomer from the acrylic, epoxy, polyester, polyether, or vinyl classes. Furthermore, for each of the aforementioned oligomer classes, the type, molecular weight, number of cross-linking sites, and glass transition temperature of the oligomer affect the performance of the coating. This results in a wide range of properties possible in each class of polyurethane coating. The performance ranges of the different types of urethanes overlap, but some broad generalization is possible. Acrylic urethanes, for example, tend to have superior resistance to sunlight, whereas polyester urethanes have better chemical resistance [1,10]. Polyurethane coatings containing polyether polyols generally have better hydrolysis resistance than acrylic- or polyester-based polyurethanes [10]. It should be emphasized that these are very broad generalizations; the performance of any specific coating depends on the particular formulation. It is entirely possible, for example, to formulate polyester polyurethanes that have excellent weathering characteristics.

The stoichiometric balance of the two reactants affects the final coating performance. Too little isocyanate can result in a soft film, with diminished chemical and weathering resistance. A slight excess of isocyanate is not generally a problem, because extra isocyanate can react with the trace amounts of moisture usually present in other components, such as pigments and solvents, or can react over time with ambient moisture. This reaction of excess isocyanate forms additional urea groups, which tend to improve film hardness. Too much excess isocyanate, however, can make the coating harder than desired, with a decrease in impact resistance. Bassner and Hegedus report that isocyanate/polyol ratios (NCO/OH) of 1.05 to 1.2 are commonly used in coating formulations to ensure that all polyol is reacted [11]. Unreacted polyol can plasticize the film, reducing hardness and chemical resistance.

2.2.3.3 Blocked Polyisocyanates

An interesting variation of urethane technology is that of the blocked polyisocyanates. These are used when chemical-cure urethane chemistry is desired but, for technical or economical reasons, a two-pack coating is not an option. Heat is needed for deblocking the isocyanate, so these coatings are suitable for use in workshops and plants, rather than in the field.

Creation of the general chemical composition consists of two steps:

1. Heat is used to deblock the isocyanate.
2. The isocyanate crosslinks with the hydrogen-containing coreactant (see Figure 2.7).

An example of the application of blocked polyisocyanate technology is polyurethane powder coatings. These coatings typically consist of a solid, blocked isocyanate and a solid polyester resin, melt blended with pigments and additives, extruded and then pulverized. The block polyisocyanate technique can also be used to formulate waterborne polyurethane coatings [8].

Additional information on the chemistry of blocked polyisocyanates is available in reviews by Potter et al. and Wicks [13-15].

2.2.3.4 Health Issues

Overexposure to polyisocyanates can irritate the eyes, nose, throat, skin, and lungs. It can cause lung damage and a reduction in lung function. Skin and respiratory

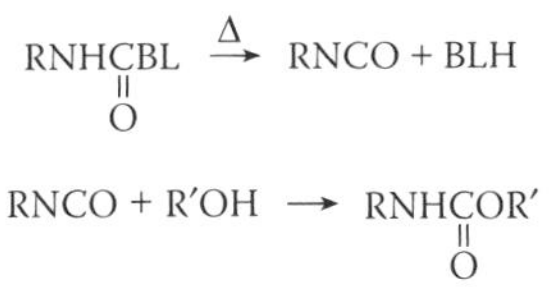

$$RNHCBL \xrightarrow{\Delta} RNCO + BLH$$
$$\qquad\ \ \overset{\|}{O}$$

$$RNCO + R'OH \longrightarrow RNHCOR'$$
$$\qquad\qquad\qquad\qquad\quad \overset{\|}{O}$$

FIGURE 2.7 General reaction for blocked isocyanates.

sensitization resulting from overexposure can result in asthmatic symptoms that may be permanent. Workers must be properly protected when mixing and applying polyurethanes as well as when cleaning up after paint application. Inhalation, skin contact, and eye contact must be avoided. The polyurethane coating supplier should be consulted about appropriate personal protective equipment for the formulation.

2.2.4.5 Waterborne Polyurethanes

For many years, it was thought that urethane technology could not effectively be used for waterborne systems because isocyanates react with water. In the past twenty years, however, waterborne polyurethane technology has evolved tremendously, and in the past few years, two-component waterborne polyurethane systems have achieved some commercial significance.

For information on the chemistry of two-component waterborne polyurethane technology, the reader should see the review of Wicks et al. [16]. A very good review of the effects of two-component waterborne polyurethane formulation on coating properties and application is available from Bassner and Hegedus [11].

2.2.4 POLYESTERS

Polyester and vinyl ester coatings have been used since the 1960s. Their characteristics include:

- Good solvent and chemical resistance, especially acid resistance (polyesters often maintain good chemical resistance at elevated temperatures [17])
- Vulnerability to attack of the ester linkage under strongly alkaline conditions

Because polyesters can be formulated to tolerate very thick film builds, they are popular for lining applications. As thin coatings, they are commonly used for coil-coated products.

2.2.4.1 Chemistry

"Polyester" is a very broad term that encompasses both thermoplastic and thermosetting polymers. In paint formulations, only thermosetting polyesters are used. Polyesters used in coatings are formed through:

1. Condensation of an alcohol and an organic acid, forming an ester — This is the unsaturated polyester prepolymer. It is dissolved in an unsaturated monomer (usually styrene or a similar vinyl-type monomer) to form a resin.
2. Crosslinking of the polyester prepolymer using the unsaturated monomer — A peroxide catalyst is added to the resin so that a free radical addition reaction can occur, transforming the liquid resin into a solid film [17].

A wide variety of polyesters are possible, depending on the reactants chosen. The most commonly used organic acids are isophthalic acid, orthophthalic anhydride,

terephthalic acid, fumaric acid, and maleic acid. Alcohol reactants used in condensation include bisphenol A, neopentyl glycol, and propylene glycol [17]. The combinations of alcohol and organic acids used determine the mechanical and chemical
properties, thermal stability, and other characteristics of polyesters.

2.2.4.2 Saponification

In an alkali environment, the ester links in a polyester can undergo hydrolysis —
that is, the bond breaks and reforms into alcohol and acid. This reaction is not
favored in acidic or neutral environments but is favored in alkali environments
because the alkali forms a salt with the acid component of the ester. These fatty acid
salts are called *soaps,* and hence this form of polymer degradation is known as
saponification.

The extent to which a particular polyester is vulnerable to alkali attack depends
on the combination of reactants used to form the polyester prepolymer and the
unsaturated monomer with which it is crosslinked.

2.2.4.3 Fillers

Fillers are very important in polyester coatings because these resins are unusually
prone to build up of internal stresses. The stresses in cured paint films arise for two
reasons: shrinkage during cure and a high coefficient of thermal expansion.

During cure, polyester resins typically shrink a relatively high amount, 8 to 10
volume percent [17]. Once the curing film has formed multiple bonds to the substrate,
however, shrinkage can freely occur only in the direction perpendicular to the
substrate. Shrinkage is hindered in the other two directions (parallel to the surface
of the substrate), thus creating internal stress in the curing film. Fillers and rein-
forcements are used to help avoid brittleness in the cured polyester film.

Stresses also arise in polyesters due to their high coefficients of thermal expansion. Values for polyesters are in the range of 36 to 72×10^{-6} mm/mm/°C, whereas
those for steel are typically only 11×10^{-6} mm/mm/°C [17]. Fillers and reinforce-
ments are important for minimizing the stresses caused by temperature changes.

2.2.5 ALKYDS

In commercial use since 1927 [18], alkyd resins are among the most widely used
anticorrosion coatings. They are one-component air-curing paints and, therefore, are
fairly easy to use. Alkyds are relatively inexpensive. Alkyds can be formulated into
both solvent-borne and waterborne coatings.

Alkyd paints are not without disadvantages:

- After cure, they continue to react with oxygen in the atmosphere, creating
  additional crosslinking and then brittleness as the coating ages [18].
- Alkyds cannot tolerate alkali conditions; therefore, they are unsuitable for
  zinc surfaces or any surfaces where an alkali condition can be expected
  to occur, such as concrete.

- They are somewhat susceptible to UV radiation, depending on the specific resin composition [18].
- They are not suitable for immersion service because they lose adhesion to the substrate during immersion in water [18].

In addition, it should be noted that alkyd resins generally exhibit poor barrier properties against moisture vapor. Choosing an effective anticorrosion pigment is therefore important for this class of coating [1].

2.2.5.1 Chemistry

Alkyds are a form of polyester. The main acid ingredient in an alkyd is phthalic acid or its anhydride, and the main alcohol is usually glycerol [18]. Through a condensation reaction, the organic acid and the alcohol form an ester. When the reactants contain multiple alcohol and acid groups, a crosslinked polymer results from the condensation reactions [18].

2.2.5.2 Saponification

In an alkali environment, the ester links in an alkyd break down and reform into alcohol and acid, (see 2.2.4.2). The known propensity of alkyd coatings to saponify makes them unsuitable for use in alkaline environments or over alkaline surfaces. Concrete, for example, is initially highly alkaline, whereas certain metals, such as zinc, become alkaline over time due to their corrosion products.

This property of alkyds should also be taken into account when choosing pigments for the coating. Alkaline pigments such as red lead or zinc oxide can usefully react with unreacted acid groups in the alkyd, strengthening the film; however, this can also create shelf-life problems, if the coating gels before it can be applied.

2.2.5.3 Immersion Behavior

In making an alkyd resin, an excess of the alcohol reagent is commonly used, for reasons of viscosity control. Because alcohols are water-soluble, this excess alcohol means that the coating contains water-soluble material and therefore tends to absorb water and swell [18]. Therefore, alkyd coatings tend to lose chemical adhesion to the substrates when immersed in water. This process is usually reversible. As Byrnes describes it, "They behave as if they were attached to the substrate by water-soluble glue [18]". Alkyd coatings are therefore not suitable for immersion service.

2.2.5.4 Brittleness

Alkyds cure through a reaction of the unsaturated fatty acid component with oxygen in the atmosphere. Once the coating has dried, the reaction does not stop but continues to crosslink. Eventually, this leads to undesirable brittleness as the coating ages, leaving the coating more vulnerable to, for example, freeze-thaw stresses.

2.2.5.5 Darkness Degradation

Byrnes notes an interesting phenomenon in some alkyds: if left in the dark for a long time, they become soft and sticky. This reaction is most commonly seen in alkyds with high linseed oil content [18]. The reason why light is necessary for maintaining the cured film is not clear.

2.2.6 CHLORINATED RUBBER

Chlorinated rubber is commonly used for its barrier properties. It has very low moisture vapor transmission rates and also performs well under immersion conditions. General characteristics of these coatings are:

- Very good water and vapor barrier properties
- Good chemical resistance but poor solvent resistance
- Poor heat resistance
- Comparatively high levels of VOCs [1,19]
- Excellent adhesion to steel [19]

Chlorinated rubber coatings have been more popular in Europe than in North America. In both markets, however, they are disappearing due to increasing pressure to eliminate VOCs.

2.2.6.1 Chemistry

The chemistry of chlorinated rubber resin is simple: polyisoprene rubber is chlorinated to a very high content, approximately 65% [19]. It is then dissolved in solvents, typically a mixture of aromatics and aliphatics, such as xylene or VM&P naphtha [19]. Because of the high molecular weight of the polymers used, large amounts of solvent are needed. Chlorinated rubber coatings have low solids contents, in the 15% to 25% (vol/vol) range.

Chlorinated rubber coatings are not crosslinked; the resin undergoes no chemical reaction during cure [1]; they are cured by solvent evaporation; in effect, the film is formed by precipitation. However, the chlorine on the rubber molecule undergoes hydrogen bonding. The tight bonding of these secondary forces gives the coating very low moisture and oxygen transmission properties.

Because the film is formed by precipitation, chlorinated rubber coatings are very vulnerable to attack by the solvents used in their formulation and have poor resistance to nearly all other solvents. They are also vulnerable to attach by organic carboxylic acids, such as acetic and formic acids [19].

2.2.6.2 Dehydrochlorination

Chlorinated rubber resins tend to undergo dehydrochlorination; that is, a hydrogen atom on one segment of the polymer molecule joins with a chlorine atom on an adjacent segment to form hydrogen chloride. When they split off from the polymer molecule, a double bond forms in their place. In the presence of heat and light, this double bond can crosslink, leading to film embrittlement. The hydrogen chloride also is a problem; in the presence of moisture, it is a source of chloride ions, which of course can initiate corrosion. The hydrogen chloride can also catalyze further breakdown of the resin [19].

Dehydrochlorination is increased by exposure to heat and light. Therefore, chlorinated rubber coatings are not suitable for use in high-temperature applications. Sensitivity to light, however, can be nullified by pigmentation.

2.2.7 OTHER BINDERS

Other types of binders include epoxy esters and silicon-based inorganic zinc-rich coatings.

2.2.7.1 Epoxy esters

Despite their name, epoxy esters are not really epoxies. Appleman, in fact, writes that epoxy esters "are best described as an epoxy-modified alkyd [20]." They are made by mixing an epoxy resin with either an oil (drying or vegetable) or a drying oil acid. The epoxy resin does not crosslink in the manner of conventional epoxies. Instead, the resin and oil or drying oil acid are subjected to high temperature, 240°C to 260°C and an inert atmosphere to induce an esterification reaction. The result is a binder that cures by oxidation and can therefore be formulated into one-component paints.

Epoxy esters generally possess adhesion, chemical and UV resistance, and corrosion protection properties that are somewhere between those of alkyds and epoxies [21]. They also exhibit resistance to splashing of gasoline and other petroleum fuels and are therefore commonly used to paint machinery [18].

2.2.7.2 Silicon-Based Inorganic Zinc-Rich Coatings

Silicon-based inorganic zinc-rich coatings are almost entirely zinc pigment; zinc levels of 90% or higher are common. They contain only enough binder to keep the zinc particles in electrical contact with the substrate and each other. The binder in inorganic ZRPs is an inorganic silicate, which may be either a solvent-based, partly hydrolyzed alkyl silicate (typically ethyl silicate) or a water-based, highly-alkali silicate.

General characteristics of these coatings are:

- Ability to tolerate higher temperatures than organic coatings (inorganic ZRPs typically tolerate 700° to 750°F)
- Excellent corrosion protection
- Require topcoatings in high pH or low pH conditions
- Require a very thorough abrasive cleaning of the steel substrate, typically near-white metal (SSPC grade SP10)

For a more-detailed discussion of inorganic ZRPs, see Section 2.3.5, "Zinc Dust."

2.3 CORROSION-PROTECTIVE PIGMENTS

2.3.1 Types of Pigments

Pigments come in three major types: inhibitive, sacrificial, and barrier. Coatings utilizing inhibitive pigments release a soluble species, such as molybdates or phosphates, from the pigment into any water that penetrates the coating. These species are carried to the metal surface, where they inhibit corrosion by encouraging the growth of protective surface layers [22]. Solubility and reactivity are critical parameters for inhibitive pigments; a great deal of research is occupied with controlling the former and decreasing the latter. Sacrificial pigments require zinc in large enough quantities to allow the flow of electric current. When in electrical contact with the steel surface, the zinc film acts as the anode of a large corrosion cell and protects the steel cathode. Both inhibitive and sacrificial pigments are effective only in the layer immediately adjacent to the steel (i.e., the primer). Barrier coatings are probably the oldest type of coating [22] and the requirements of their pigments are completely different. Specifically, chemical inertness and a flake- or plate-like shape are the requirements of barrier pigments. Unlike inhibitive or sacrificial coatings, barrier coatings can be used as primer, intermediate coat, or topcoat because their pigments do not react with metal.

2.3.1.1 A Note on Pigment Safety

The toxicity of lead, chromium, cadmium, and barium has made the continued use of paints containing these elements highly undesirable. The health and environmental problems associated with these heavy metals are serious, and new problems are discovered all the time. To address this issue, pigment manufacturers have developed many alternative pigments, such as zinc phosphates, calcium ferrites, and aluminum triphosphates, to name a few. The number of proposed alternatives is not lacking; in fact, the number and types available are nearly overwhelming.

   This chapter explores the major classes of pigments currently available for anticorrosion coating. The alert reader will quickly note that lead and barium are described here, although use of these elements can no longer be recommended. This discussion is included for two reasons. First, the protective mechanism of red lead is highly relevant to evaluating new pigments because new pigments are inevitably compared to lead. Second, the toxicity of soluble barium is less widely known than the toxicities of lead, chromium, and cadmium; therefore, barium is included here to point out that it should be avoided.

2.3.2 Lead-Based Paint

The inhibitive mechanism of the red lead found in lead-based paint (LBP) is complex. Lead pigments may be thought of as indirect inhibitors because, although they themselves are not inhibitive, they undergo a reaction with select resin systems and this reaction can form by-products that are active inhibitors [23].

2.3.2.1 Mechanism on Clean (New) Steel

Appleby and Mayne [24,25] have shown that formation of lead soaps is the mechanism used for protecting clean (or new) steel. When formulated with linseed oil, lead reacts with components of the oil to form soaps in the dry film; these soaps degrade to, among other things, the water-soluble salts lead of a variety of mono- and di-basic aliphatic acids [26,27]. Mayne and van Rooyen also showed that the lead salts of azelaic, suberic, and pelargonic acid were inhibitors of the iron corrosion. Appleby and Mayne have suggested that these acids inhibit corrosion by bringing about the formation of insoluble ferric salts, which reinforce the air-formed oxide film until it becomes impermeable to ferrous ions. This finding was based on experiments in which pure iron was immersed in a lead azelate solution, with the thickness of the oxide film measured before and after immersion. They found that the oxide film increased 7% to 17% in thickness upon immersion [25,28].

The lead salt of azelaic acid dissociates in water into a lead ion and an azelate ion. To determine which element was the key in corrosion inhibition, Appleby and Mayne also repeated the experiment with calcium azelate and sodium azelate [24,132]. Interestingly, they did not see a similar thickening of the oxide film when iron was immersed in calcium azelate and sodium azelate solutions, demonstrating that lead itself — not just the organic acid — plays a role in protecting the iron. The authors note that 5 to 20 ppm lead azelate in water is enough to prevent attack of pure iron immersed in the solution. They note that, at this low concentration, inhibition cannot be caused by the repair of the air-formed oxide film by the formation of a complex azelate, as is the case in more concentrated solutions; rather, it appears to be associated with the thickening of the air-formed oxide film. "It seems possible that, initially, lead ions in solution may provide an alternative cathodic reaction to oxygen reduction, and then, on being reduced to metallic lead at the cathodic areas on the iron surface, depolarize the oxygen reduction reaction, thus keeping the current density sufficiently high to maintain ferric film formation. In addition any hydrogen peroxide so produced may assist in keeping the iron ions in the oxide film in the ferric condition, so that thickening of the air-formed film takes place until it becomes impervious to iron ions" [25].

2.3.2.2 Mechanism on Rusted Steel

Protecting rusted steel, rather than clean or new steel, may demand of a paint a different corrosion mechanism, simply because the paint is not applied directly to the steel that must be protected but rather to the rust on top of it. Inhibitive pigments in the paint that require intimate contact with the metallic surface in order to protect it may therefore not perform well when a layer of rust prevents that immediate contact. Red-lead paint, however, does perform well on rusted steel. Several theories about the protective mechanism of red-lead paint on rusted steel exist.

2.3.2.2.1 Rust Impregnation Theory

According to this theory, the low viscosity of the vehicle used in LBP allows it to penetrate the surface texture of rust. This would have several advantages:

- Impregnation of the rust means that it is isolated and thereby inhibited in its corroding effect.

- Oil-based penetrants provide a barrier effect, thus screening the rust from water and oxygen and slowing down corrosion [29].
- Good penetration and wetting of the rust by the paint results in better adhesion.

Thomas examined cross-sections of LBP and other paints on rusted steel using transmission electron microscopy [30,31]; she found that, although the paint penetrated well into cracks in the rust layer, there was no evidence that the LBP penetrated through the compact rust layers to the rust-metal interface. (It should be noted that this experiment used cooked linseed oil, not raw; Thomas notes that raw linseed oil has a lower viscosity and might have penetrated further.) Where lead was found, it was always in the vicinity of the paint-rust interface, and in low concentrations. It had presumably diffused into the rust layer after dissolution or breakdown of the red lead pigment and was not present as discrete particles of Pb_3O_4. Thomas also found that the penetration of LBP into the rust layer wasn't significantly better than that of the other vehicles studied, for example, aluminium epoxy mastic. Finally, the penetration rate of water through linseed-oil based LBP was found to be approximately 214 $g/m^2/day$ for a 25-micron film and that of oxygen was 734 $cc/m^2/day$ for a 100-micron film [30]. The amounts of water and oxygen available through the paint film are greater than the minimum needed for the corrosion of uncoated steel. Therefore, barrier properties can be safely eliminated as the protective mechanism. Superior penetration and wetting do not appear to be the mechanisms by which LBP protects rusted steel.

2.3.2.2.2 Insolubilization of Sulfate and Chloride Theory

LBP may protect rusty steel by insolubilizing sulfate and chloride, rendering these aggressive ions inert. Soluble ferrous salts are converted into stable, insoluble, and harmless compounds; for example, sulphate nests can be rendered "harmless" by treatment with barium salts because barium sulphate is extremely insoluble. This was suggested as a protective mechanism of LBP by Lincke and Mahn [32] because, when red-lead pigmented films were soaked in concentrated solutions of Fe(II) sulfate, Fe(III) sulfate, and Fe(III) chloride, precipitation reactions occurred. Thomas [33,34] tested this theory by examining cross-sections of LBP on rusted steel (after 3 years' exposure of the coated samples) using laser microprobe mass spectrometry (LAMMS) and transmission electron microscopy with energy dispersive x-ray. Low levels of lead were found in the rust layer, but only within 30 μm of the rust-paint interface. Lead was neither seen at or near the rust-metal interface, where sulfate nests are known to exist, nor was it distributed throughout the rust layer, even though sulfur was. If rendering inert is truly the mechanism, $PbSO_4$ would be formed as the insoluble "precipitate" within the film, and the ratio of Pb to S would be 1.0 or greater (assuming a surplus of lead exists). However, no correlation was seen between the distribution of lead and that of sulfur (confirmed as sulfate by x-ray photo-electron spectroscopy); the ratio of lead to sulfur was 0.2 to 1.0, which Thomas concludes is insufficient to protect the steel. Sulfate insolubilization does not, therefore, seem to be the mechanism by which LBP protects rusted steel.

2.3.2.2.3   Cathodic Inhibition Theory

In the previously described work, low levels of lead were found in the rust layer near the paint-rust interface, within 30 μm of the rust-paint interface. Thomas suggests that because lead salts do not appear to reach the metal substrate to inhibit the anodic reaction, it is possible that lead acts within the rust layer to slow down atmospheric corrosion by interfering with the cathodic reaction (i.e., by inhibiting the cathodic reduction of existing rust [principally $FeOOH$ to magnetite]) [33]. This presumably would suppress the anodic dissolution of iron because that reaction ought to be balanced by the cathodic reaction. No conclusive proof for or against this theory has been offered.

2.3.2.2.4   Lead Soap/Lead Azelate Theory

Thomas looked for lead (as a constituent of lead azelate) at the steel-rust interface in an attempt to confirm this theory. Samples coated with lead-based paint were exposed for three years and then cross-sections were examined in a LAMMS; however, lead was not detected at the interface. As Thomas points out, this finding does not eliminate the mechanism as a possibility; lead could still be present but in levels below the 100 ppm detection limit of the LAMMS [30,31]. Appleby and Mayne have shown that 5 to 20 ppm of lead azelate is enough to protect pure iron [25]. The levels needed to protect rusted steel would not be expected to be so low, because the critical concentration required for anodic inhibitors is higher when chloride or sulphate ions are present than when used on new or clean steel [35]. Possibly, a level between 20 and 100 ppm of lead azelate is sufficient to protect the steel. Another point worth considering is that the amounts of lead that would exist in the passive film formed by complex azelates, suggested by Appleby and Mayne, has not been determined. The lead soaps/lead azelate theory appears to be the most likely mechanism to explain how red-lead paints protect rusted steel.

2.3.2.3   Summary of Mechanism Studies

Formation of lead soaps appears to be the mechanism by which lead-based paints inhibit corrosion of clean steel. When formulated with linseed oil, lead reacts with components of the oil to form soaps in the cured film; in the presence of water and oxygen, these soaps degrade to, among other things, salts of a variety of mono- and di-basic aliphatic acids. The lead salts of azelaic, suberic, and pelargonic acid act as corrosion inhibitors; lead azelate is of particular importance in LBP. These acids may inhibit corrosion by bringing about the formation of insoluble ferric salts, which reinforce the metal's oxide film until it becomes impermeable to ferrous ions, thus suppressing the corrosion mechanism.

   The formation of lead soaps is believed to be the critical corrosion-protection step for both new (clean) steel and rusted steel.

2.3.2.4   Lead-Based Paint and Cathodic Potential

Chen et al. tested red lead in an alkyd binder in both open circuit conditions and under cathodic protection. They found that this coating gave excellent service in open circuit conditions, with almost no corrosion and minimal blistering. At -1000 mV Standard

Calomel electrode (SCE), however, the same coating performed disastrously, with massive blistering and disbonding (but no corrosion). The alkyd binder with no pigment at all performed better when cathodically polarized. They suggest that, at the cathodic potential, metallic lead is deposited on the steel surface from the lead soaps. When oxygen is reduced on this lead, it produces peroxides and radicals, which the authors suggested caused disbonding at the paint-metal interface [36].

2.3.3 PHOSPHATES

"Phosphates" is a term that is used to refer to a large group of pigments that contain a phosphorus and an oxygen functional group. Its meaning is vast: the term "zinc phosphates" alone includes, but is not limited to:

- Zinc phosphate, first generation $Zn_3(PO_4)_2 \bullet 4H_2O$
- Aluminum zinc phosphate [37] or zinc aluminum phosphate [38]
- Zinc molybdenum phosphate
- Aluminium-zinc hydroxyphosphate [38]
- Zinc hydroxymolybdate-phosphate or basic zinc molybdate-phosphate [38,39]
- Basic zinc phosphate $Zn_2(OH)PO_4 \bullet 2H_2O$ [38,39]
- Zinc silicophosphate [23]
- Zinc aluminum polyphosphate [38]

Zinc-free phosphates include:

- Aluminum phosphate
- Dihydrogen tripolyphosphates [39]
- Dihydrogen aluminium triphosphate [23,37,39,40]
- Strontium aluminum polyphosphate [38]
- Calcium aluminum polyphosphate silicate [38]
- Zinc calcium strontium polyphosphate silicate [38]
- Laurylammonium phosphate [41]
- Hydroxyphosphates of iron, barium, chromium, cadmium, and magnesium, for example, $FePO_4 \bullet 2H_2O$, $Ca_3(PO_4)_2$-$1/2H_2O$, $Ba_3(PO_4)_2$, $BaHPO_4$, and $FeNH_4PO_4 \bullet 2H_2O$ [37]

In this section, the pigments discussed in more detail include the zinc phosphates and one type of nonzinc phosphate, aluminium triphosphates.

2.3.3.1 Zinc Phosphates

Zinc phosphates are widely used in many binders, including oil-based binders, alkyds, and epoxies [41–50]. Their low solubility and activity make them extremely versatile; they can be used in resins, such as alkyds, where more alkali pigments pose stability problems. Typical loading levels are 10% to 30% in maintenance coatings.

TABLE 2.3
Chronic Toxicity Data for Various Pigment Groups

Red lead	Zinc chromates	Strontium chromates	Zinc phosphates and zinc-free phosphates
Accumulation of lead, irreversible effects on hemoglobin biosynthesis, teratogenic Cat. 1	Cancero-genic Cat. 1	Cancero-genic Cat. 1	No effects observed

Source: Krieg, S., *Pitture e Vernici*, 72, 18, 1996.

The popularity of zinc phosphates — a term that encompasses an entire group of pigments — is easily understood when the toxicological data are examined. Lead, chromium, barium, and strontium are all labeled toxic in one form or another. Zinc phosphates, however, pose no known chronic toxicity. (See Table 2.3.)

The use of zinc phosphates does evoke some concerns. For example, they have shown a susceptibility to fungi attack, according to at least one researcher [51], possibly due to the nutritious properties of phosphate. In addition, Meyer has pointed out that zinc phosphate should not be used alone for longer exposure times because it hydrolyzes itself and continuously disappears from the paint film [44]; therefore, it should be used in conjunction with another anticorrosion pigment.

2.3.3.1.1 Protection Mechanism

The family of pigments known as *zinc phosphates* can provide corrosion protection to steel through multiple mechanisms:

- Phosphate Ion Donation
  Phosphate ion donation can be used for ferrous metals only [23,37,39,45, 52]. As water penetrates through the coating, slight hydrolysis of zinc phosphate occurs, resulting in secondary phosphate ions. These phosphate ions in turn form a protective passive layer [53,54] that, when sufficiently thick, prevents anodic corrosion [55]. Porosity of the phosphate coatings is closely related to the coating protective performance [37]. The approximate formula for the phosphatized metallic compound is:

$$Zn_5Fe(PO_4)2 \bullet 4H_2O \text{ [56]}.$$

- Creation of Protective Films on the Anode
  In this model, suggested by Pryor and others [57,58], oxygen dissolved in the film is adsorbed onto the metal. There it undergoes a heterogeneous reaction to form a protective film of γ–Fe_2O_3; this film thickens until it

reaches an equilibrium value of 20 nm. The film prevents the outward diffusion of iron. Phosphate ions do not appear to directly contribute to the oxide film formation but rather act to complete or maintain it by plugging discontinuities with anion precipitates of Fe(III) ions. Romagnoli has noted that Pryor used soluble phosphates rather than the generally insoluble phosphates used in coatings, so care should be taken in extrapolating these results [37]. Other studies have found both oxyhydroxides and iron phosphates incorporated in the protective film [59].

- Inhibitive Aqueous Extracts Formed with Certain Oleoresinous Binders
  Inhibitive aqueous extracts form with certain oleoresinous binders. Components of the binder, such as carboxylic and hydroxyl groups, form complexes with either the zinc phosphate or the intermediate compounds formed when the zinc phosphate becomes hydrated and dissociates. These complexes can then react with corrosion products to form a tightly-adhering, inhibitive layer on the substrate [21,39,43–46,52].

- Polarization of the Substrate
  Clay and Cox [60] have suggested that nearly insoluble basic salts are formed and adhere well to the metal surface. These salts limit the access of dissolved oxygen to the metal surface and polarize the cathodic areas. This theory was confirmed by the work of Szklarska-Smialowska and Mankowsky [61].

2.3.3.2 Types of Zinc Phosphates

Because so many variations of zinc phosphates are available, it is convenient to divide them into groups for discussion. Although no formal classes of zinc phosphates exist, they have been divided here into groups or generations, more or less by chronological development.

2.3.3.2.1 First Generation

The simplest, or first generation, zinc phosphate is made by either mixing disodium phosphate and zinc sulfate solutions at boiling temperature or saturating a 68% phosphoric acid solution with zinc oxide, also at boiling temperature. Both methods give a precipitate with an extremely coarse crystalline structure. Further treatment yields $Zn_3(PO_4)_2 \bullet 4H_2O$, first generation zinc phosphate [37].

The usefulness of first-generation zinc phosphate is limited by its low solubility [62]; only a small concentration of phosphate ions is available to protect the metal. This is a problem because corrosion inhibition by phosphates takes place only when the anion concentration is higher than 0.001M in a salt solution at pH 5.5 to 7.0 [57].

2.3.3.2.2 Second Generation

Zinc phosphates can be modified to increase their solubility in water or to add other functional groups that can also act as inhibitors. This is usually achieved by adding an organic surface treatment to the pigment or blending other inorganic inhibitors with the zinc phosphate [23]. Table 2.4 shows the amount of phosphate ions in milligrams-per-liter water obtained from various first-generation and subsequent generation zinc phosphates [63]. It can be clearly seen why modifying phosphate

TABLE 2.4
Relative Solubilities in Water of Zinc Phosphate and Modified Zinc Phosphate Pigments

| | Water-soluble matter (mg/l) (ASTM D 2448-73, 10 g pigment in 90 ml water) | | | |
Pigment	Total	Zn^{+2}	PO_4^{-3}	MoO_4^{-2}
Zinc phosphate	40	5	1	--
Organic modified zinc phosphate	300	80	1	--
Aluminum zinc phosphate	400	80	250	--
Zinc molybdenum phosphate	200	40	0.3	17

Source: Bittner, A., *J. Coat. Technol.*, Vol. 61, No. 777, p. 111, Table 2, with permission.

pigments is an area of great interest: aluminum zinc phosphate provides 250 times the amount of dissolved phosphate as first-generation zinc phosphate.

Second-generation zinc phosphates can be divided into three groups: basic zinc phosphate, salts of phosphoric acid and metallic cations, and orthophosphates.

First-generation zinc phosphate, $Zn_3(PO_4)_2 \cdot 4H_2O$, is a neutral salt. Basic zinc phosphate, $Zn_2(OH)PO_4 \cdot 2H_2O$, yields a different ratio of Zn^{2+} and PO_4^{3-} ions in solution and has a higher activity than the neutral salt [39]. It has been reported that basic zinc phosphate is as effective a corrosion inhibitor as zinc phosphate plus a mixture of pigments containing water-soluble chromates [64–66].

Another group of second-generation phosphate pigments includes salts formed between phosphoric acid and different metallic cations, for example, hydrated modified aluminium-zinc hydroxyphosphate and hydrated zinc hydroxymolybdate phosphate. Trials using these salts in alkyd binders indicate that pigments of this type can provide corrosion protection comparable to that of zinc yellow [67–69].

Orthophosphates, the third type of second-generation zinc phosphates, are prepared by reacting orthophosphoric acid with alkaline compounds [38]. This group includes:

- Zinc aluminum phosphate. It is formed by combining zinc phosphate and aluminum phosphate in the wet phase; the aluminum ions hydrolyze, causing acidity, which in turn increases the phosphate concentration [38,70,71]. Aluminum phosphate is added to give higher phosphate content.
- Organically modified basic zinc phosphates. An organic component is fixed onto the surface of basic zinc phosphate particles, apparently to improve compatibility with alkyd and physically drying resins.
- Basic zinc molybdenum phosphate hydrate. Zinc molybdate is added to basic zinc phosphate hydrate so it can be used with water-soluble systems,

for example, styrene-modified acrylic dispersions [38]. The pigment produces a molybdate anion (MoO_4^{-2}) that is an effective anodic inhibitor; its passivating capacity is only slightly less than that of the chromate anion [37].

2.3.3.2.3 Third Generation

The third generation of zinc phosphates consists of polyphosphates and polyphosphate silicates. Polyphosphates — molecules of more than one phosphorous atom together with oxygen — result from condensation of acid phosphates at higher temperatures than used to produce orthophosphates [38]. This group includes:

- Zinc aluminum polyphosphate. This pigment contains a higher percentage of phosphate, as P_2O_5, than zinc phosphate or modified zinc orthophosphates.
- Strontium aluminum polyphosphate. This pigment also has greater phosphate content than first-generation zinc phosphate. The solubility behavior is further altered by inclusion of a metal whose oxides react basic compared to amphoteric zinc [38].
- Calcium aluminum polyphosphate silicate. This pigment exhibits an altered solubility behavior due to calcium. The composition is interesting: active components are fixed on the surface of an inert filler, wollastonite.
- Zinc calcium strontium polyphosphate silicate. In this pigment, the electrochemically active compounds are also fixed on the surface of wollastonite.

2.3.3.3 Accelerated Testing and Why Zinc Phosphates Commonly Fail

Although zinc phosphates show acceptable performance in the field, they commonly show inferior performance in accelerated testing. This response is probably affected by their very low solubility. In accelerated tests, the penetration rate of aggressive ions is highly speeded up, but the solubility of zinc phosphate is not. The amount of aggressive ions thus exceeds the protective capacity of both the phosphate anion and the iron oxide layer on the metal substrate [37]. Bettan has postulated that there is an initial lag time with zinc phosphates because the protective phosphate complex forms slowly on steel's surface. Because the amount of corrosion-initiating ions is increased from the very beginning of an accelerated test, corrosion processes can be initiated during this lag time. In field exposure, lag time is not a problem, because the penetration of aggressive species usually also has its own lag time. Angelmayer has supported this explanation also [66,72].

Romagnoli [37] also points out that researcher findings conflict and offers some possible reasons why:

- Experimental variables of the zinc phosphate pigments may differ. One example is distribution of particle diameter; smaller diameter means increased surface area, which increases the amount of phosphate leaching from the pigment. The more phosphate anion in a solution, the better the anticorrosion protection. Pigment volume concentration (PVC) and critical PVC (CPVC) for the particular paint formulations used are also important and frequently neglected. And, of course, because the term *zinc phosphate* applies to both a family of pigments and a specific formula, the exact type of zinc phosphate is important.

*   Binder type and additives are not the same. In accelerated testing, the type of binder is usually the most important factor because of its barrier properties. Only after the binder barrier is breached does effect of pigment become apparent.

2.3.3.4 Aluminum Triphosphate

Hydrated dihydrogen aluminium triphosphate ($AlH_2P_3O_{10} \bullet 2H_2O$) is an acid with a dissociation constant, pKa, of approximately 1.5 to 1.6. Its acidity per unit mass is approximately 10 to 100 times higher than other similar acids, such as aluminium and silicon hydroxides.

When dissolved, aluminium triphosphate dissociates into triphosphate ions:

$$AlH_2P_3O_{10} \rightarrow Al^{3+} + 2H^+ + [P_3O_{10}]^{5-}$$

Beland suggests that corrosion protection comes both from the ability of the tripolyphosphate ion to chelate iron ions (passivating the metal) and from tripolyphosphate ions' ability to depolymerize into orthophosphate ions, giving higher phosphate levels than zinc or molybdate phosphate pigments [23].

Chromy and Kaminska attribute the corrosion protection entirely to the triphosphate. They suggest that the anion $(P_3O_{10})^{5-}$ reacts with anodic iron to yield an insoluble layer, which is mainly ferric triphosphate. This phosphate coating is insoluble in water, is very hard, and exhibits excellent adhesion to the substrate [39].

Aluminum triphosphate has limited solubility in water and is frequently modified with either zinc or silicon to control both solubility and reactivity [23,29]. Researchers have demonstrated that aluminium triphosphate is compatible with various binders, including long-, medium-, and short-oil alkyds; epoxies; epoxy-polyesters; and acrylic-melamine resins [73–76]. Chromy notes that it is particularly effective on rapidly corroding coatings; it may therefore be useful in overcoating applications [39].

Nakano has found that aluminium triphosphate can outperform zinc chromate and calcium plumbate pigments in a chlorinated rubber vehicle. Testing in this study involved only salt spray, no field exposure. The substrate was galvanized steel, and the pigments were used in both chlorinated rubber and an air-drying alkyd. Aluminium triphosphate performed better in the chlorinated rubber [74]. Noguchi has seen that aluminium triphosphate in an alkyd vehicle performed better than zinc chromate and zinc phosphate, again using salt spray testing only [77].

2.3.3.5 Other Phosphates

Phosphate pigments other than zinc and aluminium phosphates have received much less attention in the technical literature. This group includes phosphates, hydroxyphosphates, and acid phosphates of the metals iron, barium, chromium, cadmium, and magnesium. For iron and barium, the only important phosphates appear to be $FePO_4 \cdot 2H_2O$, $Ca_3(PO_4)_2 - 1/2H_2O$, $Ba_3(PO_4)_2$, $BaHPO_4$, and $FeNH_4PO_4 \cdot 2H_2O$ [37, 39]. Iron phosphate by itself gives poor results, at least in accelerated testing, but appears promising when used with basic zinc phosphate. Reaction accelerators, such as sodium molybdate and sodium m-nitrobenzene sulphonate, have been found to improve the corrosion resistance of coatings containing iron phosphate [78].

Calcium acid phosphate, $CaHPO_4$, has also been discussed in the literature as an anticorrosion pigment. Vetere and Romagnoli have studied it as a replacement for zinc tetroxychromate. When used in a phenolic chlorinated rubber binder, calcium acid phosphate outperformed the simplest zinc phosphate [$Zn_3(PO_4)_2$] and was comparable to zinc tetroxychromate in salt spray testing. However, researchers were not able to identify the mechanism by which this pigment could offer protection to metal. Iron samples in an aqueous suspension of the pigment showed some passivity in corrosion potential measurements. Analysis of the protective layer's composition showed that it is composed mostly of iron oxides; calcium and phophate ions are present but not, perhaps, at the levels expected for a good passivating pigment [79].

Another phosphate pigment that has been studied is lauryl ammonium phosphate. However, very little information is available about this pigment. Gibson briefly describes studies using lauryl ammonium phosphate, but the results do not seem to warrant further work with this pigment [41].

2.3.4 FERRITES

Ferrite pigments have the general formula $MeO \cdot Fe_2O_3$, where Me = Mg, Ca, Sr, Ba, Fe, Zn, or Mn. They are manufactured by calcination of metal oxides. The principal reaction is:

$$MeO + Fe_2O_3 \rightarrow MeFe_2O_4$$

at temperatures of approximately 1000°C. These high temperatures translate into high production costs for this class of pigments [23].

Ferrite pigments appear to protect steel both by creating an alkaline environment at the coating-metal interface and, with certain binders, by forming metal soaps. Kresse [70,80] has found that zinc and calcium ferrites react with fatty acids in the binder to form soaps and attributes the corrosion protection to passivation of the metal by the alkaline environment thus created in the coating.

Sekine and Kato [81] agree with this soap formation mechanism. However, they have also tested several ferrite pigments in an epoxy binder, which is not expected to form soaps with metal ions. All of the ferrite-pigmented epoxy coatings offered better corrosion protection than both the same binder with red iron oxide as anticorrosion pigment and the binder with no anticorrosion pigment. Examination of the rest potential versus immersion time of the coated panels showed a lag time between initial immersion and passivation of approximately 160 hours in this study. The authors concluded that passivation of the metal occurs only after water has permeated the coating and reached the paint or metal interface [82]. The delay in onset of passivation could perhaps also be explained if, as in LBP, the protection

TABLE 2.5
Corrosion Rate of Mild Steel in Extracted
Aqueous Solution of Pigments

Pigment	pH	Corrosion rate, mg/dm^2/day
Mg ferrite	8.82	12.75
Ca ferrite	12.35	0.26
Sr ferrite	7.85	16.71
Ba ferrite	8.20	18.00
Fe ferrite	8.40	14.95
Zn ferrite	7.31	14.71
Red iron oxide	3.35	20.35
No Pigment	6.15	15.82

Reprinted with permission from: Sekine, I. and Kato, T., *Ind. Eng. Chem. Prod. Res. Dev.*, 25, 7, 1986. Copyright 1986, American Chemistry Society.

mechanism depends on a breakdown of the soaps and passivation is achieved with a soap degradation product.

Sekine and Kato also examined the pH of aqueous extractions of ferrite pigments and the corrosion rate of mild steel immersed in these solutions [82]. Their results are presented in Table 2.5. These data are interesting because they imply that, in addition to soap formation, the pigments can also create an alkali environment at the metal or paint interface. These authors have found that the corrosion-protective properties of the ferrite pigments in epoxy paint films, based on electrochemical measurements, were (in decreasing order) Mg>Fe>Sr>Ca>Zn>Ba. It should be emphasized that this ranking was obtained in one study: the relative ranking within the ferrite group may owe much to such variables as particle size of the various pigments and pigment volume concentration (comparable percent weights rather than PVC were used).

Verma and Chakraborty [83] compared zinc ferrite and calcium ferrite to red lead and zinc chromate pigments in aggressive industrial environments. The vehicle used for the pigments was a long oil linseed alkyd resin. Panels were exposed for eight months in five fertilizer plant environments: a urea plant, an ammonium nitrate plant, a nitrogen-phosphorous-potassium (NPK) plant, a sulfuric acid plant, and a nitric acid plant where, the authors note, acid fumes and fertilizer dust spills are almost continual occurrences. Results vary greatly, depending on plant type. In the sulfuric acid plant, the two ferrites outperformed the lead and chromate pigments by a very wide margin. In the urea and NPK plants, the calcium ferrite pigment was better than any other pigment. In the ammonium nitrate plant, the calcium ferrite pigment performed substantially worse than the others. In the nitric acid plant, the zinc chromate pigment performed significantly worse than the other three, but among these three, the difference was not substantial. The authors attribute the superior behavior of calcium ferrite over zinc ferrite to the former's controlled but higher solubility. Metal ions in solution, they suggest, react with aggressive species that are permeating into the coating and thus prevent them from reaching the metal-coating interface.

An interesting aspect of the ferrites is that their corrosion-protection mechanism, and the binders with which they can be used, are very similar to that of red lead pigment. These pigments may be of particular interest, therefore, in overcoating aged LBP. A major requirement of successful overcoating is compatibility between the old coating and the new coating; this is greatly enhanced by using the same binder type in both.

2.3.5 ZINC DUST

Zinc-rich paints (ZRPs) are, of course, not new; they have been used to protect steel construction for many decades [84]. Zinc dust comes in two forms: the normally used and highly effective flake zinc dust and the less-expensive granular grade. The difference between flake zinc dust and the less-effective granular grade is important; Zimmerman has experimented with replacing part of the flake grade with granular zinc dust and found that, when the amount of flake fell below 25% of dry coating weight (that is, 1/3 of the total pigment), performance was very poor. It was possible, however, to somewhat reduce the amount of flake zinc dust by replacing it with granular zinc dust or micaceous iron oxide (MIO) and still obtain good coating performance [85].

Zinc dust offers corrosion protection to steel via four mechanisms:

1. Cathodic protection to the steel substrate (the zinc acts as a sacrificial anode). This takes place at the beginning of the coating's lifetime and naturally disappears with time [86].
2. Barrier action. As a result of the zinc sacrificially corroding, zinc ions are released into the coating. These ions can react with other species in the coating to form insoluble zinc salts. As they precipitate, these salts fill in the pores in the coating, reducing permeability of the film [84].
3. Oxygen reduction. Molecular oxygen diffusing through the coating toward the metal is consumed in a reaction with metallic zinc. The zinc particles form a layer of ZnO and $Zn(OH)_2$; de Lame and Piens have found that the rate of oxygen reduction decreases exponentially with an increase in the thickness of this layer. They speculate that the mechanism of oxygen reduction could last longer than that of cathodic protection [87].
4. Slightly alkaline conditions are formed as the zinc corrodes [86]. For this reason, of course, only binders that tolerate some degree of alkalinity must be used.

Of these four mechanisms, the first two depend on a high zinc content to work properly; the last two are independent of zinc content.

There are two types of ZRPs, which differ depending on the binder used: organic and inorganic. Two-component epoxy amine or amides, epoxy esters, and moisture-cure urethanes are examples of organic binders. Organic binders have a dense character and are electrically insulating; for that reason, the PVC/CPVC ratio must be greater than 1 for the zinc to perform as a sacrificial anode. This requirement — the reverse of what is usually seen in the coatings world — is necessary to ensure electrical conductivity. If the PVC is less than the CPVC, the zinc particles are not in direct electrical contact with each other, and the insulating binder between the particles prevents the bulk of the zinc dust from offering cathodic protection to the steel.

Inorganic binders are silica-based. They can be further divided into two groups: solvent-based partly hydrolyzed alkyl silicate (mostly ethyl silicate) and water-based highly alkaline silicates. Inorganic ZRPs are conductive and are therefore used as weldable or shop primers. They also have high porosities. With time (and corrosion of the zinc), the matrix fills with zinc salts, giving a very dense barrier coat. Inorganic ethyl silicate in partly hydrolyzed form sometimes has a storage stability problem.

Inorganic ZRPs require higher film builds than do the organic ZRPs. Schmid recommends approximately 50 μm with an organic one-component binder, approximately 75 μm with an organic two-component binder, and approximately 100 μm with an inorganic binder [88]. Other workers in the field have proposed film builds of up to 140 μm for inorganic binders.

2.3.6 CHROMATES

The chromate passivating ion is among the most efficient passivators known. However, due to health and environmental concerns associated with hexavalent chromium, this class of anticorrosion pigments is rapidly disappearing.

2.3.6.1 Protection Mechanism

Simply put, chromate pigments stimulate the formation of passive layers on metal surfaces [89]. The actual mechanism is probably more complex. Svoboda has described the protection mechanism of chromates as "a process which begins with physical adsorption which is transformed to chemisorption and leads to the formation of compounds which also contain trivalent chromium" [90].

In the mechanism described by Rosenfeld et al. [91], CrO_4^{2-} groups are adsorbed onto the steel surface, where they are reduced to trivalent ions. These trivalent ions participate in the formation of the complex compound $FeCr_2O_{14-n}(OH^-)_n$, which in turn forms a protective film. Largin and Rosenfeld have proposed that chromates do not merely form a mixed oxide film at the metal surface; instead, they cause a change in the structure of the existing oxide film, accompanied by a considerable increase in the bond energy between the iron and oxygen atoms. This leads to an increase in the protective properties of the film [92].

It should perhaps also be noted that several workers in the field describe the protection mechanism more simply as the formation of a normal protective mixed oxide film, with defects in the film plugged by Cr_2O_3 [23,57].

2.3.6.2 Types of Chromate Pigments

The principal chromate-based pigments are basic zinc potassium chromate (also known as *zinc yellow* or *zinc chrome*), strontium chromate, and zinc tetroxychromate. Other chromate pigments exist, such as barium chromate, barium potassium chromate, basic magnesium chromate, calcium chromate, and ammonium dichromate; however, because they are used to a much lesser extent, they are not discussed here.

Zinc potassium chromate is the product of inhibitive reactions among potassium dichromate, zinc oxide, and sulfuric acid. Zinc chromates are effective inhibitors even at relatively low loading levels [23].

Strontium chromate, the most expensive chromate inhibitor, is mainly used on aluminium. It is used in the aviation and coil-coating industries because of its effectiveness at very low loadings.

Zinc tetroxychromate, or basic zinc chromate, is commonly used in the manufacture of two-package polyvinyl butyryl (PVB) wash primers. These consist of phosphoric acid and zinc tetroxychromate dispersed in a solution of PVB in alcohol. These etch primers, as they are known, are used to passivate steel, galvanized steel, and aluminium surfaces, improving the adhesion of subsequent coatings. They tend to be low in solids and are applied at fairly low film thicknesses [23].

2.3.6.3 Solubility Concerns

The ability of a chromate pigment to protect a metal lies in its ability to dissolve and release chromate ions. Controlling the solubility of the pigment is critical for chromates. If the solubility is too high, other coating properties, such as blister formation, are adversely affected. A coating that uses a highly soluble chromate pigment under long-term moisture conditions can act as a semipermeable membrane: with water on one side (at the top of the coating) and a saturated solution of aqueous pigment extract on the other (at the steel-coating interface). Significant osmotic forces thus lead to blister formation [90]. Chromate pigments are therefore not suitable for use in immersion conditions or conditions with long periods of condensation or other moisture exposure.

2.3.7 OTHER INHIBITIVE PIGMENTS

Other types of inhibitive pigments include calcium-exchanged silica, barium metaborate, molybdates, and silicates.

2.3.7.1 Calcium-Exchanged Silica

Calcium-exchanged silica is prepared by ion-exchanging an anticorrosion cation, calcium, onto the surface of a porous inorganic oxide of silica. The protection mechanism is ion-exchange: aggressive cations (e.g., H^+) are preferentially exchanged onto the pigment's matrix as they permeate the coating, while Ca^{2+} ions are simultaneously released to protect the metal. Calcium does not itself passivate the metal or otherwise directly inhibit corrosion. Instead, it acts as a flocculating agent. The small amounts (circa 120 µm /ml H_2O at pH ≈ 9) of silica in solution flocculate around the Ca^{2+} ion. The Ca–Si species has a small δ+ or δ– charge, which drives it toward the metal surface (due to the potential drop across the metal/solution interface). Particles of silica and calcium agglomerate at the paint/metal interface. There the alkaline pH causes spontaneous coalescing into a thin film of silica and calcium [93]. The major benefit of this inorganic film seems to be that it prevents Cl^- and other corrosion-initiating species from reaching the metal surface.

The dual action of entrapment of aggressive cations and release of inhibitor gives calcium-exchange silica two advantages over traditional anticorrosion pigments:

1. The "inhibitor" ion is only released in the presence of aggressive cations, which means that no excess of the pigment to allow for solubility is necessary.
2. No voids are created in the film by the ion-exchange; the coating has fairly constant permeability characteristics [38,93–95].

2.3.7.2 Barium Metaborate

Barium metaborate is a pigment to avoid. It contains a high level of soluble barium, an acute toxicant. Disposal of any waste containing this pigment is likely to be expensive, whether that waste is produced in the manufacture or application of the coatings or much later when preparing to repaint structures originally coated with barium metaborate.

Barium metaborate creates an alkaline environment, inhibiting the steel; the metaborate ion also provides anodic passivation [23]. The pigment requires high loading levels, up to 40% of coating weight, according to Beland. It is highly soluble and fairly reactive with several kinds of binders; this leads to stability problems when formulated with such products as acidic resins, high-acid number resins, and acid-catalyzed baking systems. A modified silica coating is often used to reduce and control solubility. One way to decrease its reactivity and, therefore, increase the number of binders with which it can be used, is to modify it with zinc oxide or a combination of zinc oxide and calcium sulphate [23]. The high loading level required for heavy-duty applications implies that careful attention must be paid to the PVC/CPVC ratio when formulating with this pigment.

Information regarding actual service performance of barium metaborate coatings is scarce, and what does exist does not seem to justify the use of this pigment. In the early 1980s, the state of Massachusetts repair-painted a bridge with barium metaborate pigment in a conventional oil/alkyd vehicle. The result was not satisfactory: after six years, considerable corrosion had occurred at the beam ends and on the railings above the road [22]. It should perhaps be noted that an alkyd vehicle is not the ideal choice for a pigment that generates an alkaline environment; better results may perhaps have been obtained with a higher-performance binder. However, because of the toxicity problems associated with soluble barium, further work with barium metaborate does not seem to be warranted.

2.3.7.3 Molybdates

Molybdate pigments are calcium or zinc salts precipitated onto an inert core such as calcium carbonate [47,96–98]. They prevent corrosion by inhibiting the anodic corrosion reaction [47]. The protective layer of ferric molybdate, which these pigments form on the surface of the steel, is insoluble in neutral and basic solutions.

Use of these pigments has been limited because of their expense. Zinc phosphate versions of the molybdate pigments have been introduced in order to lower costs and improve both adhesion to steel and film flexibility [23,47,96–98]. The molybdate pigment family includes:

- Basic zinc molybdate
- Basic calcium zinc molybdate
- Basic zinc molybdate/phosphate
- Basic calcium zinc molybdate/zinc phosphate

In general, tests of these pigments as corrosion inhibitors in paint formulations have returned mixed results on steel. Workers in the field tend to refer somewhat wistfully to the possibilities of improving the performance of molybdates through combination with other pigments, in the hope of obtaining a synergistic effect. A serious drawback is that, in several studies, molybdates appeared to cause coating embrittlement, perhaps due to premature binder aging [99–102].

Although molybdate pigments are considered nontoxic [103], they are not completely harmless. When cutting or welding molybdate-pigmented coatings, fumes of very low toxicity are produced. With proper ventilation, these fumes are not likely to prove hazardous [101]. The possible toxicity is about 10% to 20% that of chromium compounds [103,104].

2.3.7.4 Silicates

Silicate pigments consist of soluble metallic salts of borosilicates and phosphosilicates. The metals used in silicate pigments are barium, calcium, strontium, and zinc; silicates containing barium can be assumed to pose toxicity problems.

The silicate pigments include:

- Calcium borosilicates, which are available in several grades, with varying B_2O_3 content (not suitable for immersion or semi-immersion service or water-based resins [23])
- Calcium barium phosphosilicate
- Calcium strontium phosphosilicate
- Calcium strontium zinc phosphosilicate, which is the most versatile phosphosilicate inhibitor in terms of binder compatibility [23]

The silicate pigments can inhibit corrosion in two ways: through their alkalinity and, in oleoresinous binders, by forming metal soaps with certain components of the vehicle. Which process predominates is not entirely clear, perhaps because the efficacy of the pigments is not entirely clear. When Heyes and Mayne examined calcium phosphosilicate and calcium borosilicate pigments in drying oils, they found a mechanism similar to that of red lead: the pigment and the oil binder react to form metal soaps, which degrade and yield products with soluble, inhibitive anions [105].

Van Ooij and Groot found that calcium borosilicate worked well in a polyester binder, but not in an epoxy or polyurethane [106]. This hints that the alkalinity generated within the binder cannot be very high, otherwise the polyester — being much more vulnerable to saponification — would have shown much worse results than either the epoxy or the polyurethane. Metal soaps, of course, would not be formed with either an epoxy or polyurethane. However, the possibility of metal soaps cannot be absolutely ruled out for a polyester without knowing exactly what is meant by this unfortunately broad term.

The state of Massachusetts had a less-positive experience with the same pigment, although possibly a different grade of it. In the 1980s, the state of Massachusetts repair-painted a number of bridges with calcium borosilicate pigment in a conventional ole-oresinous binder — a vehicle that would presumably form metal soaps. Spot blasting was performed prior to coating. The calcium borosilicate system was judged less forgiving of poor surface preparation than is LBP, and attaining the minimum film build was found to be critical. Massachusetts eventually stopped using this pigment because of the high costs of improved surface preparation and inspection of film build [0].

Another silicate, calcium barium phosphosilicate, has been tested in conjunction with six other pigments on cold-rolled steel in an epoxy-polyamide binder [0, 0]. After nine months' atmospheric exposure in a marine environment (Biarritz, France), the samples with calcium barium phosphosilicate pigment — and those with barium metaborate — gave worse results than either the aluminum triphosphate or ion-exchanged calcium silicate pigments. (These in turn were significantly outperformed by a modified zinc phosphate as well as by zinc chromate pigment.)

2.3.8 BARRIER PIGMENTS

2.3.8.1 Mechanism and General Information

Barrier coatings protect steel against corrosion by reducing the permeability of liquids and gases through a paint film. How much the permeability of water and oxygen can be reduced depends on many factors, including:

- Thickness of the film
- Structure of the film (polymer type used as binder)
- Degree of binder crosslinking
- Pigment volume concentrations
- Type and particle shape of pigments and fillers

Pigments used for barrier coatings are diametrically opposed to the active pig-ments used in other anticorrosion coatings in one respect: in barrier coatings, they must be inert and completely insoluble in water. Commonly used barrier pigments can be broken into two groups:

- Mineral–based materials, such as mica, MIO, and glass flakes
- Metallic flakes of aluminium, zinc, stainless steel, nickel, and cupronickel

In the second group, care must be taken to avoid possible electrochemical interac-tions between the metallic pigments and the metal substrate [109].

2.3.8.2 Micaceous Iron Oxide

MIO is a naturally occurring iron oxide pigment that contains at least 85% Fe_2O_3. The term "micaceous" refers to its particle shape, which is flake-like or lamellar: particles are very thin compared to their area. This particle shape is extremely important for MIO in protecting steel. MIO particles orient themselves within the coating, so that the flakes are lying parallel to the substrate's surface. Multiple layers of flakes form an effective barrier against moisture and gases [40,109–116]. MIO is fascinating in one respect: it is a form of rust that has been used as an effective pigment in barrier coatings for decades to protect steel from ... rusting.

For effective barrier properties, PVCs in the range of 25% to 45% are used, and the purity must be at least 80% MIO (by weight). Because MIO is a naturally occurring mineral, it can vary from source to source, both in chemical composition and in particle size distribution. Smaller flakes mean more layers of pigment in the dried film, which increases the pathway that water must travel to reach the metal. Schmid has noted that, in a typical particle-size distribution, as much as 10% of the particles may be too large to be effective in thin coatings, because there are not enough layers of flakes to provide a barrier against water. To provide a good barrier in the vicinity of these large particles, MIO is used in thick coatings or multicoats [88].

Historically, it has been believed that MIO coatings tend to fail at sharp edges because the miox particles were randomly oriented in the vicinity of edges. Random orientation would, of course, increase the capillary flow of water along the pigment's surface toward the metal substrate. However, Wiktorek and Bradley examined coverage over sharp edges using scanning electron microscope images of cross-sections. They found that lamellar miox particles always lie parallel to the substrate, even over sharp edges. The authors suggested that when failure is seen at edges, the problem is really thinner coatings in these areas [117].

In addition to providing a barrier against diffusion of aggressive species through the coating, MIO confers other advantages:

- It provides mechanical reinforcement to the paint film.
- It can block ultraviolet light, thus shielding the binder from this destructive form of radiation.

For the latter reason, MIO is sometimes used in topcoat formulations to improve weatherability [40,109].

The chemical inertness of MIO means that it can be used in a variety of binders: alkyd, chlorinated rubber, styrene-acrylic and vinyl copolymers, epoxy, and polyurethane [40].

2.3.8.2.1 Interactions of MIO with Aluminum

It is not clear from the literature whether or not combining MIO and aluminum pigments in a coating poses a problem. There are recommendations both for and against mixing MIO with these pigments.

In full-scale trials of various paint systems on bridges in England, Bishop found that topcoats with both MIO and aluminum pigments form a white deposit over large areas. Analysis showed these deposits to be mostly aluminum sulphate with some ammonium sulphate. The only possible source of aluminum in the coating system was the topcoat pigment. Bishop did not find the specific cause of this problem. He notes that bridge paints in the United states commonly contain leafing aluminum and that few problems are reported [118].

Schmid, on the other hand, recommends combining MIO with other lamellar materials, such as aluminum flake and talc, to improve the barrier properties of the film by closer pigment packing [88].

2.3.8.3 Other Nonmetallic Barrier Pigments

2.3.8.3.1 Mica

Mica is a group of hydrous potassium aluminosilicates. The diameter-to-thickness ratio of this group exceeds 25:1, higher than that of any other flaky pigment. This makes mica very effective at building up layers of pigment in the dried film, thus increasing the pathway that water must travel to reach the metal and reducing water permeability [119,220].

2.3.8.3.2 Glass

Glass fillers include flakes, beads, microspheres, fibers, and powder. Glass flakes provide the best coating barrier properties. Other glass fillers can also form a protective barrier because of their close packing in the paint coating. Glass has been used in the United States, Japan, and Europe when high-temperature resistance, or high resistance to abrasion, erosion, and impact, is needed. The thicknesses of coatings filled with glass flakes are approximately 1 to 3 mm; flakes are 3 to 5 μm thick, so every millimeter of coating can contain approximately 100 layers of flakes [109].

Studies have shown that glass flakes perform comparably to lamellar pigments of stainless steel and MIO pigments but perform worse than aluminum flake; the latter showed better flake orientation than glass flake in the paint film [109,121–123]. Glass flake is usually preferred for elevated temperatures, not only because of its ability to maintain chemical resistance at high temperatures but also because of its coefficient of thermal expansion. Coatings filled with glass flake can obtain thermal expansion properties close to those of carbon steel. This enables them to retain good adhesion even under thermal shock [124,125].

Glass beads, microspheres, fibers, and powders are also used for their thermal properties in fire-resistant coatings. Spherical glass beads can increase the mechanical strength of a cured film. Using beads of various diameters can improve packing inside the dry film, thus improving barrier properties. Glass fibers impart good abrasion resistance to the paint. Glass microspheres are a component of the fly ash produced by the electric power industry. More precisely, they are aluminosilicate spheres, with diameters between 0.3 and 200 μm, that are composed of Al_2O_3, Fe_2O_3, CaO, MgO, Na_2O, and K_2O. The exact makeup depends on the type and source of fuel burned [109].

2.3.8.4 Metallic Barrier Pigments

2.3.8.4.1 Aluminum

Besides reducing the permeability of water vapor, oxygen, and other corrosive media, aluminum pigment also reflects UV radiation and can withstand elevated temperatures.

There are two types of aluminum pigment: leafing and nonleafing. Leafing pigment orients itself parallel to the substrate at the top of the coating; this positioning enables the pigment to protect the binder against UV damage but may not be the best location for maximizing barrier properties. Leafing properties depend on the presence of a thin fatty acid layer, commonly stearic acid, on the flakes. Nonleafing aluminum pigments have a more random orientation in the coating and are very effective in barrier coatings [109]. De and colleagues, for example, have obtained favorable results with aluminum in a chlorinated rubber vehicle in seawater trials in India [126].

2.3.8.4.2   Zinc Flakes

Zinc flakes should not to be confused with the zinc dust used in zinc-rich coatings: the size is of a different magnitude altogether. Some research suggests that zinc flakes could give both the cathodic protection typical of zinc dust and the barrier protection characteristic of lamellar pigments [109]. However, in practice, this could be very difficult to achieve because the zinc dust particles in zinc-rich paints have to be in electrical contact to obtain cathodic protection. Designing a coating in which the zinc particles are in intimate contact with each other and with the steel, and yet completely free of gaps between pigment and binder or between pigment particles, is difficult. The lack of any gaps is critical for a barrier pigment, because it is precisely these gaps that provide the easy route for water and oxygen to reach the metal surface. In fact, Hare and Wright's [127] research shows that zinc flakes undergo rapid dissolution in corrosive environments when they are used as the sole pigment in paints; their coatings are prone to blistering.

2.3.8.4.3   Other Metallic Pigments

Other metallic pigments, such as stainless steel, nickel, and copper, have also been used in recent years. Their use in coatings of metals with more noble electrochemical potential than carbon steel entails a certain risk of galvanic corrosion between the coating and the substrate. The pigment volume concentrations in such paints must be kept well below the levels at which the metallic pigment particles are in electrical contact with each other and the carbon steel. If this condition is not met, pitting follows. Bieganska recommends using a nonconducting primer as an insulating layer between the steel substrate and the barrier coating, if it is necessary to use a strong electropositive pigment in the barrier layer [109]. The same author also warns that, although the mechanical durability and high-temperature resistance of stainless steel flake makes this type of pigment desirable, it is not suited to applications where chlorides are present [109].

Nickel flake-filled coatings can be useful for strongly alkaline environments. Cupronickel flakes (Cu – 10% Ni – 2% Sn) are used in ship protection because of their outstanding antifouling properties. The alloy pigment is of interest in this application because its resistance to leaching is better than that of copper itself [109].

2.3.9   CHOOSING A PIGMENT

Before choosing a pigment and formulating paint, one question must be answered: will an active or a passive role be required of the pigment? The role of the pigment — active or passive — must be decided at the start for the fairly straightforward reason that only one or the other is possible. Many of the pigments that actively inhibit corrosion, such as through passivation, must dissolve into anions and cations; ion species can then passivate the metal surface. Without water, these pigments do not dissolve and the protection mechanism is not triggered. And, of course, it is the express purpose of barrier coatings to prevent water from reaching the coating-metal interface.

Once the role of the pigment has been decided, choice of pigment depends on such factors as:

- *Price.* Many of the newer pigments are expensive. The amounts necessary in a coating, and the respective impact on price, plays a large role in determining whether the pigment is economically feasible.
- *Commercial availability.* Producing a few hundred grams of a pigment in a laboratory is one thing; however, it is quite another to generate a pigment in hundreds of kilograms for commercial paints.
- *Difficulty of blending into a real formulation.* Pigments must do more than just protect steel. They have to disperse in the wet paint, rather than stay clumped together. They also have to be well attached to the binder so that water cannot penetrate through the coating via gaps between pigment particles and the binder. In many cases, the surfaces of pigments are chemically treated to avoid these problems; however, it must be possible to treat pigments without changing their essential properties (solubility, etc.).
- *Suitability in the binders that are of interest.* A coating does not, of course, consist merely of a pigment; the binder is of equal importance in determining the success of a paint. The pigments chosen for further study must be compatible with the binders of interest.
- Resistance to heat, acids or alkalis, and/or solvents, as needed.

2.4 ADDITIVES

For corrosion-protective purposes, the most important components of a coating are the binder and the anticorrosion pigment. Additives are necessary for the manufacture, application, and cure of a coating; however, with the exception of corrosion inhibitors, they play a relatively minor role in corrosion protection.

This section presents a brief overview of some of the additives found in modern anticorrosion coatings. The field of coating composition is too complex to be covered in any depth in the following sections and, in any case, numerous texts devoted to the science — or art — of coating formulation already exist.

2.4.1 FLOW AND DISPERSION CONTROLLERS

Flow and dispersion controllers are used to control the behavior of the wet paint, either in the paint can during mixing and application or between application and cure. This group of additives includes thixotropic agents, surfactants, dispersants, and antiflooding/antifloating agents. Thixotropic agents and surfactants are the most important of the flow and dispersion controllers.

2.4.1.1 Thixotropic Agents

Thixotropic agents are used to control the rheology of a coating — that is, how thick the coating is under various conditions, how much it spreads, and how quickly it does so. This group includes several very different types of additives: thickeners, antisagging compounds, antisettling and suspension agents, antigelling agents, leveling and coalescing aids, wet-edge extenders, anticratering agents, and plasticizers [128].

The rheology of a coating might need to be modified for a number of reasons. One is to prevent pigment sedimentation; the pigment must be able to remain in suspension after mixing, rather than settling in a solid mass at the bottom of the container before it can be applied. Rheology is also modified so that the coating can be applied in a particular method. Brush, roller, spray, curtain, and knife coating techniques all produce different amounts of shear in the paint at the moment it contacts the substrate. For a thixotropic coating, in which viscosity is inversely related to shear, this means that the coating will have very different viscosities at the moment of application and hence different wetting and spreading behaviors. Rheology modifiers are used to control the shear viscosity for the various application methods, so that the coating wets and spreads on the metal surface [3].

Examples of thixotropic agents include fumed silicas and treated clays. These inert pigments are sometimes added to aid in film build, to add body to a paint, or for antisettling characteristics [2].

2.4.1.2 Surfactants

Surfactants are used when the surface energy of a coating as a whole, or one or more of its components, must be controlled. This group of additives includes wetting agents, pigment dispersers, defoamers, and antifoaming agents. Wetting agents help lower the surface tension of the coating, so that it spreads out and wets the surface, forming a continuous film across it.

Foaming problems can occur both in the manufacture of the coating and in its application. Defoamers are used to prevent such problems, especially in waterborne formulations [3]. "Microfoaming" is the term for the tiny bubbles that occasionally form on the surface of a wet film, affecting the film appearance. They are more commonly seen in waterborne coatings than in solvent-borne ones and can be prevented with antifoaming agents.

2.4.1.3 Dispersing Agents

Pigments are generally manufactured to a specific particle size, or range of sizes, for optimal strength and opacity (if the pigment is a filler), color strength (if the pigment is a colorant), solubility rate (anticorrosion pigments), and other desired properties. However, during transportation and storage, pigment particles tend to agglomerate. In the process of making paint, these agglomerations must be broken up and the pigment or coating must be treated with an additive to ensure that the pigment particles stay dispersed. This additive is known as the *dispersing agent.* In solvent-borne paints, the dispersing agent is commonly a steric barrier, whereas in waterborne coatings, electrostatic repulsion is used [29].

2.4.2 Reactive Reagents

Reactive reagents generally aid in film formation, forming bonds to the substrate, crosslinking, and curing. Examples of this class of additives include metallic driers, such as zinc or tin salts, to aid in crosslinking [10,18]; curing catalysts and accelerators; photoinitiators; and adhesion promoters.

2.4.3 Contra-Environmental Chemicals

As their name implies, contra-environmental chemicals are a group of additives that are intended to provide the coating with protection against its service environment. Examples of this type of additive include [128]:

- Performance enhancers (antiskinning agents, antioxidants, light stabilizers, nonpigmental corrosion inhibitors)
- Thermal controllers (freeze-thaw controllers, heat stabilizers)
- Biological controllers (biocides, antifouling agents)

Antioxidants and light stabilizers   are used to provide topcoats with thermo-oxidative and UV stabilization, thus increasing service life in outdoor applications. For thermo-oxidative stabilization, phenolic antioxidants and aromatic amines are generally used [129]. Hindered amine light stabilizers (HALS; for example, Hostavin N30™, Goodrite 3150™, Chimassorb 944™) or UV absorbers (for example, Cyasorb UV-531™) [130] are added to the coating mostly for UV protection and, to some extent, for thermo-oxidative stabilization. A mixture of antioxidants and light stabilizers is frequently used; this must be carefully formulated because both positive and negative effects have been reported from combining these additives [131,132]. Barret and colleagues suggest that the phenol in the antioxidant prevents the conversion of HALS to a stabilizing nitroxide [133]. Another mechanism may be that the radicals of different stabilizers interact.

The term **corrosion inhibitors** is not meant to include anticorrosion pigments in this section. These additives are completely soluble in order to provide the maximum possible corrosion protection immediately upon application of the paint. Pigments have a much more controlled solubility rate in order to have an effect over a long period. Corrosion inhibitors are commonly used for preventing spot or "flash" rusting. Sodium nitrate, for example, is sometimes added to waterborne coatings to prevent flash rusting [3]. These corrosive-inhibiting additives are used in addition to, rather than as a substitute for, anticorrosion pigments. Corrosion inhibitors and anticorrosion pigments must be chosen with care if used together, so as not to adversely affect the in-can stability of the formulation [3].

Biocides prevent microbial growth in coatings, both in-can and in the cured paint. They are more important in waterborne coatings than in solvent-borne coatings.

Antifouling agents   prevent the growth of mussels, sea urchins, and other marine life on marine coatings. They are used exclusively in topcoats, rather than in the primers that provide the corrosion protection to the metal substrate.

2.4.4 SPECIAL EFFECT INDUCERS

Special effect inducers are additives that are used to help the coating meet special or unusual requirements. Examples include:

- Surface conditioners (gloss controllers, texturing agents)
- Olfactory controllers (odorants and deodorants)

REFERENCES

1. Smith, L.M., *J. Prot. Coat. Linings,* 13, 73, 1995.
2. Salem, L.S., *J. Prot. Coat. Linings,* 13, 77, 1996.
3. Flynn, R. and Watson, D., *J. Prot. Coat. Linings,* 12, 81, 1995.
4. Bentley, J., Organic film formers, in *Paint and Surface Coatings Theory and Practice,* Lambourne, R., Ed., Ellis Horwood Limited, Chichester, 1987.
5. Forsgren, A., Linder, M. and Steihed, N., *Substrate-polymer compatibility for various waterborne paint resins,* Report 1999:1E, Swedish Corrosion Institute, Stockholm, 1999.
6. Billmeyer, F.W., *Textbook of Polymer Science,* 3rd ed., John Wiley & Sons, New York, 1984, 388.
7. Brendley, W.H., *Paint Varnish Prod.,* 63, 19, 1973.
8. Potter, T.A. and Williams, J.L., *J. Coat. Technol.,* 59, 63, 1987.
9. Gardner, G., *J. Prot. Coat. Linings,* 13, 81, 1996.
10. Roesler, R.R. and Hergenrother, P.R., *J. Prot. Coat. Linings,* 13, 83, 1996.
11. Bassner, S. L. and Hegedus, C.R., *J. Prot. Coat. Linings,* 13, 52, 1996.
12. Luthra, S. and Hergenrother, R., *J. Prot. Coat. Linings,* 10, 31, 1993.
13. Potter, T.A., Rosthauser, J.W. and Schmelzer, H.G., in *Proc., 11th International Conference Organic Coatings Science Technology,* Athens, 1985, Paper 331.
14. Wicks, Z.W. Jr., *Prog. Org. Coat.,* 9, 3, 1981.
15. Wicks, Z.W. Jr., *Prog. Org. Coat.,* 3, 73, 1975.
16. Wicks, Z.W. Jr., Wicks, D.A. and Rosthauser, J.W., *Prog. Org. Coat.,* 44, 161, 2002.
17. Slama, W.R., *J. Prot. Coat. Linings,* 13, 88, 1996.
18. Byrnes, G., *J. Prot. Coat. Linings,* 13, 73, 1996.
19. Hare, C.H., *J. Prot. Coat. Linings,* 12, 41, 1995.
20. Appleman, B.R., *Corrosioneering,* 1, 4, 2001.
21. Kaminski, W., *J. Prot. Coat. Linings,* 13, 57, 1996.
22. Hare, C.H., *Mod. Paint Coat.,* 76, 38, 1986.
23. Beland, M., *Am. Paint Coat. J.,* 6, 43, 1991.
24. Appleby, A.J. and Mayne, J.E.O., *JOCCA,* 59, 69, 1976.
25. Appleby, A.J. and Mayne, J.E.O., *JOCCA,* 50, 897, 1967.
26. Mayne, J.E.O. and van Rooyen, D., *J. Appl. Chem.,* 4, 419, 1960.
27. Mayne, J.E.O. and Ramshaw, E.H., *J. Appl. Chem.,* 13, 553, 1963.
28. Hancock, P., *Chemistry and Industry,* 194, 1961.
29. van Oeteren, K.A. *Farben-Chem.,* 73, 12, 1971.
30. Thomas, N.L., *Prog. Org. Coat.,* 19, 101, 1991.
31. Thomas, N.L., *Proc. Symp. Advances in Corrosion Protection by Organic Coatings,* Electrochem. Soc., 1989, 451.

32. Lincke, G. and Mahn, W.D., *Proc. 12th FATIPEC Congress, Fédération d'Associations de Techniciens des Industries des Peintures, Vernis, Emaux et Encres d'Imprimerie de l'Europe Continentale (FATIPEC),* Paris, 1974, 563.
33. Thomas, N.L., in *Proc. PRA Symposium, Coatings for Difficult Surfaces,* Hampton (UK), 1990, Paper No. 10.
34. Thomas, N.L., *J. Prot. Coat. and Linings,* 6, 63, 1989.
35. Brasher, D.M. and Mercer, A.D., *Brit. Corros. J.,* 3, 120, 1968.
36. Chen, D., Scantlebury, J.D. and Wu, C.M., *Corros. Mat.,* 21, 14, 1996.
37. Romagnoli, R. and Vetere, V.F. *Corros. Rev.,* 13, 45, 1995.
38. Krieg, S., *Pitture e Vernici,* 72, 18, 1996.
39. Chromy, L. and Kaminska, E., *Prog. Org. Coat.,* 18, 319, 1990.
40. Boxall, J., *Polym. Paint Colour J.,* 179, 127, 1989.
41. Gibson, M.C. and Camina, M., *Polym. Paint Colour J.,*178, 232, 1988.
42. Ruf, J., *Werkst. Korros.,* 20, 861, 1969.
43. Meyer, G., *Farbe+Lack,* 68, 315, 1962.
44. Meyer, G., *Farbe+Lack,* 69, 528, 1963.
45. Meyer, G., *Farbe+Lack,* 71, 113, 1965.
46. Meyer, G., *Werkst. Korros.,* 16, 508, 1963.
47. Boxall, J., *Paint & Resin,* 55, 38, 1985.
48. Ginsburg, T., *J. Coat. Technol.,* 53, 23, 1981.
49. Gomaa, A.Z. and Gad, H.A., *JOCCA,* 71, 50, 1988.
50. Svoboda, M., *Farbe+Lack,* 92, 701, 1986.
51. Stranger-Johannessen, M., *Proc., 18th FATIPEC Congress (Vol. 3), Fédération d'Associations de Techniciens des Industries des Peintures, Vernis, Emaux et Encres d'Imprimerie de l'Europe Continentale (FATIPEC),* Paris, 1987, 1.
52. Robu, C., Orban, N. and Varga, G., *Polym. Paint Colour J.,* 177, 566, 1987.
53. Bernhard, A., Bittner, A. and Gawol, M., *Eur. Suppl. Poly. Paint Colour J.,* 171, 62, 1981.
54. Ruf, J., *Chimia,* 27, 496, 1973.
55. Dean, S.W., Derby, R. and von der Bussche, G., *Mat. Performance,* 12, 47, 1981.
56. Kwiatkowski, L., Lampe, J. and Kozlowski, A., *Powloki Ochr.,* 14, 89, 1988. Summarized in Chromy, L. and Kaminska, E. *Prog. Org. Coat.,* 18, 319, 1990.
57. Leidheiser, H. Jr., *J. Coat. Technol.,* 53, 29, 1981.
58. Pryor, M.J. and Cohen, M., *J. Electrochem. Soc.,* 100, 203, 1953.
59. Kozlowski, W. and Flis, J., *Corr. Sci.,* 32, 861, 1991.
60. Clay, M.F. and Cox, J.H. *JOCCA,* 56, 13, 1973.
61. Szklarska-Smialowska, Z. and Mankowsky, J., *Br. Corros. J.,* 4, 271, 1969.
62. Burkill, J.A. and Mayne, J.E.O., *JOCCA,* 9, 273, 1988.
63. Bittner, A., *J. Coat. Technol.,* 61, 111, 1989.
64. Adrian, G., *Pitture Vernici,* 61, 27, 1985.
65. Bettan, B., *Pitture Vernici,* 63, 33, 1987.
66. Bettan, B., *Paint and Resin,* 56, 16, 1986.
67. Adrian, G., Bittner, A. and Carol, M., *Farbe+Lack,* 87, 833, 1981.
68. Adrian, G., *Polym. Paint Colour J.,* 175, 127, 1985.
69. Bittner, A., *Pitture Vernici,* 64, 23, 1988.
70. Kresse, P., *Farbe+Lack,* 83, 85, 1977.
71. Gerhard, A. and Bittner, A., *J. Coat. Technol.,* 58, 59, 1986.
72. Angelmayer, K-H., *Polym. Paint Colour J.,* 176, 233, 1986.
73. Nakano, J. et al., *Polym. Paint Colour J.,* 175, 328, 1985.
74. Nakano, J. et al., *Polym. Paint Colour J.,* 175, 704, 1985.

75. Nakano, J. et al., *Polym. Paint Colour J.*, 177, 642, 1987.
76. Takahashi, M., *Polym. Paint Colour J.*, 177, 554, 1987.
77. Noguchi, T. et al., *Polym. Paint Colour J.*, 173, 888, 1984.
78. Gorecki, G., *Metal Fin.*, 90, 27, 1992.
79. Vetere, V.F. and Romagnoli, R., *Br. Corros. J.*, 29, 115, 1994.
80. Kresse, P., *Farbe und Lacke*, 84, 156, 1978.
81. Sekine, I. and Kato, T., *JOCCA*, 70, 58, 1987.
82. Sekine, I. and Kato, T., *Ind. Eng. Chem. Prod. Res. Dev.*, 25, 7, 1986.
83. Verma, K.M. and Chakraborty, B.R., *Anti-Corrosion*, 34, 4, 1987.
84. Boxall, J., *Polym. Paint Colour J.*, 181, 443, 1991.
85. Zimmerman, K., *Eur. Coat. J.*, 1, 14 1991.
86. Piens, M., *Evaluations of Protection by Zinc Primers*, presentation seminar at Liege, Coatings Research Institute, Limelette, Oct. 25-26, 1990.
87. de Lame, C. and Piens, M., *Reactivite de la poussiere de zinc avec l'oxygene dissous*, *Proc., XXIII FATIPEC Congress, Fédération d'Associations de Techniciens des Industries des Peintures, Vernis, Emaux et Encres d'Imprimerie de l'Europe Continentale (FATIPEC)*, Paris, 1996, A29-A36.
88. Schmid, E.V., *Polym. Paint Colour J.*, 181, 302, 1991.
89. Pantzer, R., *Farbe und Lacke*, 84, 999, 1978.
90. Svoboda, M. and Mleziva, J., *Prog. Org. Coat.*, 12, 251, 1984.
91. Rosenfeld, I.L. et al., *Zashch. Met.*, 15, 349, 1979.
92. Largin, B.M. and Rosenfeld, I.L., *Zashch. Met.*, 17, 408, 1981.
93. Goldie, B.P.F., *JOCCA*, 71, 257, 1988.
94. Goldie, B.P.F., *Paint and Resin*, 1, 16, 1985
95. Goldie, B.P.F., *Polym. Paint Colour J.*, 175, 337, 1985.
96. Banke, W.J., *Mod. Paint Coat.*, 2, 45, 1980.
97. Sullivan, F.J. and Vukasovich, M.S., *Mod. Paint Coat.*, 3, 41, 1981.
98. Garnaud, M.H.L., *Polym. Paint Colour J.*, 174, 268, 1984.
99. Lapain, R., Longo, V. and Torriano, G., *JOCCA*, 58, 286, 1975.
100. Marchese, A., Papo, A. and Torriano, G., *Anti-Corrosion*, 23, 4, 1976.
101. Lapasin, R., Papo, A. and Torriano, G., *Brit. Corros. J.*, 12, 92, 1977.
102. Wilcox, G.D., Gabe, D.R. and Warwick, M.E. *Corros. Rev.*, 6, 327, 1986.
103. Sherwin-Williams Chemicals, New York, Technical Bulletin No. 342.
104. *Threshold Limit Values for Chemical Substances and Biological Exposure Indices*, Vol. 3, American Conference of Governmental Industrial Hygienists, Cincinnati, 1971, 192.
105. Heyes, P.J. and Mayne, J.E.O., in *Proc. 6th Eur. Congr. on Metallic Corros.*, London, 1977, 213.
106. van Ooij, W.J. and Groot. R.C., *JOCCA*, 69, 62, 1986.
107. Amirudin, A. et al., *Prog. Org. Coat.*, 25, 339, 1995.
108. Amirudin, A., and Thierry, D., *Brit. Corros. J.*, 30, 128, 1995.
109. Bieganska, B., Zubielewicz, M. and Smieszek, E., *Prog. Org. Coat.*, 16, 219, 1988.
110. Bishop, D.M. and Zobel, F.G., *JOCCA*, 66, 67, 1983.
111. Bishop, D.M., *JOCCA*, 64, 57, 1981.
112. Wiktorek, S. and John, J., *JOCCA*, 66, 164, 1983.
113. Boxall, J., *Polym. Paint Colour J.*, 174, 272, 1984.
114. Carter, E., *Polym. Paint Colour J.*, 171, 506, 1981.
115. Schmid, E.V., *Farbe+Lack*, 90, 759, 1984.
116. Schuler, D., *Farbe+Lack*, 92, 703, 1986.
117. Wiktorek, S. and Bradley, E.G., *JOCCA*, 7, 172, 1986

118. Bishop, R.R., *Brit. Corrosion J.,* 9, 149, 1974.
119. Various authors, in *Surface Coatings,* Vol. 1, Waldie, J.M., Ed., Chapman and Hall, London, 1983.
120. Eickhoff, A.J., *Mod. Paint Coat.,* 67, 37, 1977.
121. Hare, C.H. and Fernald, M.G., *Mod. Paint Coat.,* 74, 138, 1984.
122. Hare, C.H., *Mod. Paint Coat.,* 75, 37, 1985.
123. El-Sawy, S.M. and Ghanem, N.A., *JOCCA,* 67, 253, 1984.
124. Hearn, R.C., *Corros. Prev. Control,* 34, 10, 1987.
125. Sprecher, N., *JOCCA,* 66, 52, 1983.
126. De, C.P. et al., in *Proc. 5th Internat. Congress Marine Corros. Fouling, ASM International,* Materials Park (OH), 1980, 417.
127. Hare, C.H. and Wright, S. J., *J. Coat. Technol.,* 54, 65, 1982.
128. Verkholantsev, V., *Eur. Coat. J.,* 12, 32, 1998.
129. Schmitz, J. et al., *Prog. Org. Coat.,* 35, 191, 1999.
130. Sampers, J., *Polym. Degradation and Stability,* 76, 455, 2002.
131. Pospíšil, J, and Klemchuk, P., *Oxidation Inhibition in Organic Materials,* CRC Press, Boca Raton, Florida, 1990.
132. Rychla, L. et al., *Int. J. Polym. Mater.,* 13, 227, 1990.
133. Barret, J. et al., *Polym. Degradation and Stability,* 76, 441, 2002.

3 Waterborne Coatings

Most of the important types of modern solvent-borne coatings — epoxies, alkyds, acrylics — are also available in waterborne formulations. In recent years, even urethane polymer technology has been adapted for use in waterborne coatings [1]. However, waterborne paints are not simply solvent-borne paints in which the organic solvent has been replaced with water; the paint chemist must design an entirely new system from the ground up. In this chapter, we discuss how waterborne paints differ from their solvent-borne counterparts.

Waterborne paints are by nature more complex and more difficult to formulate than solvent-borne coatings. The extremely small group of polymers that are soluble in water does not, with a few exceptions, include any that can be usefully used in paint. In broad terms, a one-component, solvent-borne coating consists of a polymer dissolved in a suitable solvent. Film formation consists of merely applying the film and waiting for the solvent to evaporate. In a waterborne latex coating, the polymer particles are not at all dissolved; instead they exist as solid polymer particles dispersed in the water. Film formation is more complex when wetting, thermodynamics, and surface energy theory come into play. Among other challenges, the waterborne paint chemist must:

- Design a polymer reaction to take place in water so that monomer building blocks polymerize into solid polymer particles
- Find additives that can keep the solid polymer particles in a stable, even dispersion, rather than in clumps at the bottom of the paint can
- Find more additives that can somewhat soften the outer part of the solid particles, so that they flatten easier during film formation.

And all of this was just for the binder. Additional specialized additives are needed, for example, to keep the pigment from clumping; these are usually different for dispersion in a polar liquid, such as water, than in a nonpolar organic solvent. The same can be said for the chemicals added to make the pigments integrate well with the binder, so that gaps do not occur between binder and pigment particles. And, of course, more additives unique to waterborne formulations may be used to prevent flash rusting of the steel before the water has evaporated. (It should perhaps be noted that the need for flash rusting additives is somewhat questionable.)

3.1 TECHNOLOGIES FOR POLYMERS IN WATER

Most polymer chains are not polar; water, being highly polar, cannot dissolve them. Chemistry, however, has provided ways to get around this problem. Paint technology has taken several approaches to suspending or dissolving polymers in water. All of them require some modification of the polymer to make it stable in a water dispersion or solution. The concentration of the polar functional groups plays a role in deciding the form of the waterborne paint: a high concentration confers water-solubility, whereas a low concentration leads to dispersion [2]. Much research has been ongoing to see where and how polar groups can be introduced to disrupt the parent polymer as little as possible.

3.1.1 WATER-REDUCIBLE COATINGS AND WATER-SOLUBLE POLYMERS

In both water-reducible coatings and water-soluble polymers, the polymer chain, which is naturally hydrophobic, is altered; hydrophilic segments such as carboxylic acid groups, sulphonic acid groups, and tertiary amines are grafted onto the chain to confer a degree of water solubility.

In water-reducible coatings, the polymer starts out as a solution in an organic solvent that is miscible with water. Water is then added. The hydrophobic polymer separates into colloid particles, and the hydrophilic segments stabilize the colloids [3]. Water-reducible coatings, by their nature, always contain a certain fraction of organic solvent.

Water-soluble polymers do not begin in organic solvent. These polymers are designed to be dissolved directly in water. An advantage to this approach is that drying becomes a much simpler process because the coating is neither dispersion nor emulsion. In addition, temperature is not as important for the formation of a film with good integrity. The polymers that lend themselves to this technique, however, are of lower molecular weight (10^3 to 10^4) than the polymers used in dispersions (10^5 to 10^6) [4].

3.1.2 AQUEOUS EMULSION COATINGS

An emulsion is a dispersion of one liquid in another; the best-known example is milk, in which fat droplets are emulsified in water. In an emulsion coating, a liquid polymer is dispersed in water. Many alkyd and epoxy paints are examples of this type of coating.

3.1.3 AQUEOUS DISPERSION COATINGS

In a aqueous dispersion coatings, the polymer is not water–soluble at all. Rather, it exists as a dispersion or latex of very fine (50 to 500 nm diameter) solid particles in water. It should be noted that merely creating solid polymer particles in organic solvent, removing the solvent, and then adding the particles to water does not produce aqueous dispersion coatings. For these coatings, the polymers must be produced in water from the start. Most forms of latex begin as emulsions of the polymer building blocks and then undergo polymerization. Polyurethane dispersions, on the other hand, are produced by polycondensation of aqueous building blocks [3].

3.2 WATER VS. ORGANIC SOLVENTS

The difference between solvent-borne and waterborne paints is due to the unique character of water. In most properties that matter, water differs significant from organic solvents. In creating a waterborne paint, the paint chemist must start from scratch, reinventing almost everything from the resin to the last stabilizer added.

Water differs from organic solvents in many aspects. For example, its dielectric constant is more than an order of magnitude greater than those of most organic solvents. Its density, surface tension, and thermal conductivity are greater than those of most of the commonly used solvents. For its use in paint, however, the following differences between water and organic solvents are most important:

- **Water does not dissolve the polymers that are used as resins in many paints.** Consequently the polymers have to be chemically altered so that they can be used as the backbones of paints. Functional groups, such as amines, sulphonic groups, and carboxylic groups, are added to the resins to make them soluble or dispersible in water.
- **The latent heat of evaporation is much higher for water, than for organic solvents.** Thermodynamically driven evaporation of water occurs more slowly at room temperature.
- **The surface tension of water is higher than those of the solvents commonly used in paints.** This high surface tension plays an important part in the film formation of latexes (see Section 3.3).

3.3 LATEX FILM FORMATION

Waterborne dispersions form films through a fascinating process. In order for crosslinking to occur and a coherent film to be built, the solid particles in dispersion must spread out as the water evaporates. They will do so because coalescence is thermodynamically favored over individual polymer spheres: the minimization of total surface allows for a decrease in free energy [5].

Film formation can be described as a three-stage process. The stages are described below; stages 1 and 2 are depicted in Figure 3.1.

1. **Colloid concentration.** The bulk of the water in the newly applied paint evaporates. As the distance between the spherical polymer particles shrinks, the particles move and slide past each other until they are densely packed. The particles are drawn closer together by the evaporation of the water but are themselves unaffected; their shape does not change.
2. **Coalescence.** This stage begins when the only water remaining is in-between the particles. In this second stage, also called the "capillary" stage," the high surface tension of the interstitial water becames a factor. The water tries to reduce its surface at both the water-air and water-particle

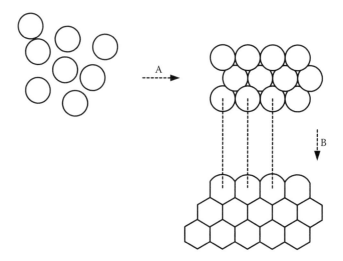

FIGURE 3.1 Latex film formation: colloid concentration (A) and coalescence (B). Note that center-to-center distances between particles do not change during coalescence.

interfaces. The water actually pulls enough on the solid polymer particles to deform them. This happens on the sides, above, and below the sphere; everywhere it contacts another sphere, the evaporating water pulls it toward the other sphere. As this happens on all sides and to all spheres, the result is a dodecahedral honeycomb structure.

3. **Macromolecule interdiffusion.** Under certain conditions, such as sufficiently high temperatures, the polymer chains can diffuse across the particle boundaries. A more homogeneous, continuous film is formed. Mechanical strength and water resistance of the film increase [5, 6].

3.3.1 DRIVING FORCE OF FILM FORMATION

The film formation process is extremely complex, and there are a number of theories — or more accurately, schools of theories — to describe it. A major point of difference among them is the driving force for particle deformation: surface tension of the polymer particles, Van der Waals attraction, polymer-water interfacial tension, capillary pressure at the air-water interface, or combinations of the above. These models of the mechanism of latex film formation are necessary in order to improve existing waterborne paints and to design the next generation. To improve the rate of film formation, for example, it is important to know if the main driving force for coalescence is located at the interface between polymer and water, between water and air, or between polymer particles. This location determines which surface tension or surface energies should be optimized.

In recent years, a consensus seems to be growing that the surface tension of water, either at the air-water or the polymer-water interface — or both — is the driving force. Atomic force microscopy (AFM) studies seem to indicate that capillary pressure at the air-water interface is most important [7]. Working from another approach, Visschers and

TABLE 3.1
Estimates of Forces Operating During Particle Deformation

Type of Force Operating	Estimated Magnitude (N)
Gravitational force on a particle	6.4×10^{-17}
Van der Waals force (separation 5 nm)	8.4×10^{-12}
Van der Waals force (separation 0.2 nm)	5.5×10^{-9}
Electrostatic repulsion	2.8×10^{-10}
Capillary force due to receding water-air interface	2.6×10^{-7}
Capillary force due to liquid bridges	1.1×10^{-7}

Reprinted from: Visschers, M., Laven, J., and Vander Linde, R., *Prog. Org. Coat.*, 31, 311, 1997. With permission from Elsevier.

colleagues [9] have reported supporting results. They estimated the various forces that operate during polymer deformation for one system, in which a force of 10^{-7} N would be required for particle deformation. The forces generated by capillary water between the particles and by the air-water interface are both large enough. (See Table 3.1.)

Gauthier and colleagues have pointed out that polymer-water interfacial tension and capillary pressure at the air-water interface are expressions of the same physical phenomenon and can be described by the Young and Laplace laws for surface energy [5]. The fact that there are two minimum film formation temperatures, one "wet" and one "dry," may be an indication that the receding polymer-water interface and evaporating interstitial water are both driving the film formation (see Section 3.4).

For more in-depth information on the film formation process and important thermodynamic and surface-energy considerations, consult the excellent reviews by Lin and Meier [7]; Gauthier, Guyot, Perez, and Sindt [5]; or Visschers, Laven, and German [9]. All of these reviews deal with nonpigmented latex systems. The reader working in this field should also become familiar with the pioneering works of Brown [10], Mason [11], and Lamprecht [12].

3.3.2 HUMIDITY AND LATEX CURE

Unlike organic solvents, water exists in the atmosphere in vast amounts. Researchers estimate that the atmosphere contains about 6×10^{15} liters of water [13,14]. Because of this fact, relative humidity is commonly believed to affect the rate of evaporation of water in waterborne paints. Trade literature commonly implies that waterborne coatings are somehow sensitive to high-humidity conditions. However, Visschers, Laven, and van der Linde have elegantly shown this belief to be wrong. They used a combination of thermodynamics and contact-angle theory to prove that latex paints dry at practically all humidities as long as they are not directly wetted — that is, by rain or condensation [8]. Their results have been borne out in experiments by Forsgren and Palmgren [15], who found that changes in relative humidity had no significant effect on the mechanical and physical properties of the cured coating. Gauthier and colleagues have also shown experimentally that latex coalescence does not depend on ambient humidity. In studies of water evaporation using weight-loss measurements, they found that the rate in stage 1 depends on ambient humidity for a given temperature. In stage 2, however, when coalescence occurs, water evaporation rate could not be explained by the same model [5].

3.3.3 REAL COATINGS

The models for film formation described above are based on latex-only systems. Real waterborne latex coatings contain much more: pigments of different kinds (see chapter 2); coalescing agents to soften the outer part of the polymer particles; and surfactants, emulsifiers, and thickeners to control wetting and viscosity and to maintain dispersion.

Whether or not a waterborne paint will succeed in forming a continuous film depends on a number of factors, including:

- Wetting of the polymer particles by water (Visschers and colleagues found that the contact angle of water on the polymer sphere has a major influence on the contact force that pushes the polymer particles apart [if positive] or pulls them together [if negative] [8])
- Polymer hardness
- Effectiveness of the coalescing agents
- Ratio of binder to pigment
- Dispersion of the polymer particles on the pigment particles
- Relative sizes of pigment to binder particles in the latex

3.3.3.1 Pigments

To work in a coating formulation, whether solvent-borne or waterborne, a pigment must be well dispersed, coated by a binder during cure, and in the proper ratio to the binder. The last point is the same for solvent-borne and waterborne formulations; however, the first two require consideration in waterborne coatings.

The high surface tension of water affects not only polymer dispersion but also pigment dispersion. As Kobayashi has pointed out, the most important factor in dispersing a pigment is the solvent's ability to wet it. Because of surface tension considerations, wetting depends on two factors: hydrophobicity (or hydrophilicity) of the pigment and the pigment geometry. The interested reader is directed to Kobayashi's review for more information on pigment dispersion in waterborne formulations [16].

Joanicot and colleagues examined what happens to the film formation process described above when pigments much larger in size than latex particles are added to the formulation. They found that waterborne formulations behave similarly to solvent-borne formulations in this matter: the pigment volume concentration (PVC) is critical. In coatings with low PVC, the film formation process is not affected by the presence of pigments. With high PVC, the latex particles are still deformed as water evaporates but do not exist in sufficent quantity to spread completely over the pigment particles. The dried coating resembles a matrix of pigment particles that are held together at many points by latex particles [17].

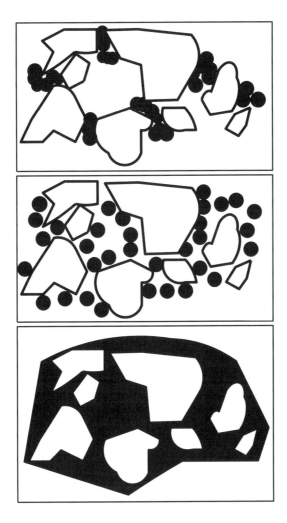

FIGURE 3.2 Pigment and binder particle combinations. The polymer particles are black, and the pigment particles are white or striped (representing two different pigments). Top: High PVC, with binder particles aggregated between pigment particles. Middle: High PVC and dispersed binder particles. Bottom: Low PVC and enough binder to fill all gaps between pigment particles.

The problems of PVC-pigment dispersion imbalance are shown in Figure 3.2.

In the top part of Figure 3.2, the PVC is very high and the binder particles have flocculated at a limited number of sites between pigment particles. When they deform, the film will consist of pigment particles held together in places by polymer, with voids throughout.

The middle section of Figure 3.2 shows the same very high PVC, but here the binder particles are dispersed. The binder particles may form a continuous film around the pigment particles, but voids still occur because there simply is not enough binder.

The bottom part of Figure 3.2 shows the ideal scenario: the PVC is lower, and the surrounding black binder is able to not only cover the pigment particles but also leave no void between them.

3.3.3.2 Additives

In real waterborne paints, the film formation process can result in a nonhomogeneous layer of cured paint. Tzitzinou and colleagues, for example, have shown that the composition of a cured paint layer can be expected to vary through the depth of the coating. They studied an anionic surfactant in an acrylic latex film. Using AFM and Rutherford backscattering spectrometry on cured films, they found a higher concentration of surfactant at the air surface than in the bulk of the coating [18]. Wegmann has also studied the inhomogeneity of waterborne films after cure, but attributes his findings mainly to insufficient coalescence during cure [19].

The chemistry of real latex formations is complex and currently defies predictive modeling. A reported problem for waterborne modelers is that an increase in curing temperature can affect various coating components differently. Snuparek and colleagues added a nonionic emulsifier to a dispersion of copolymer butyl methacrylate/butyl acrylate/acrylic acid. When cure took place at room temperature, the water resistance of the films increased with the amount of emulsifier added. When cure happened at 60°C, however, the water resistance of the films *decreased* with the amount of emulsifier added [4].

3.4 MINIMUM FILM FORMATION TEMPERATURE

Minimum film formation temperature (MFFT) is the minimum temperature needed for a binder to form a coherent film. This measurement is based on, although not identical to, the glass transition temperature (T_g) of the polymer.

If a coating is applied below the MFFT, the water evaporates as described for Stage 1 (see Section 3.3). However, because the ambient temperature is below the MFFT, the particles are too hard to deform. Particles do not coalesce as the interstitial water evaporates in stage 2. A honeycomb structure, with Van der Waals bonding between the particles and polymer molecules diffused across particle boundaries, does not occur.

The MFFT can be measured in the laboratory as the minimum temperature at which a cast latex film becomes clear. This is simply because if the coating has not formed a coherent film, it will contain many voids between polymer particles. These voids create internal surfaces within the film, which cause the opacity.

Latexes must always be applied at a temperature above the MFFT. This is more difficult than it sounds, because the MFFT is a dynamic value, changing over time. In a two-component system, the MFFT begins increasing as soon as the components are mixed. Two-component waterborne paints must be applied and dried before the MFFT has increased enough to reach room temperature. When the MFFT has reached room temperature and the end-of-pot life has been reached for a waterborne paint, viscosity does not increase as it does with many solvent-borne paints [20].

3.4.1 WET MFFT AND DRY MFFT

If a latex paint is dried below the MFFT, no particle deformation occurs. However, if the temperature of the dried (but not coalesced) latex is then raised to slightly above the MFFT, no coalescence as described in Section 3.3 should occur; no receding air-water interface exists to generate capillary forces, and thus no particle deformation occurs. If the temperature is further raised, however, particle deformation eventually occurs. This is because some residual water is always left between the particles due to capillary condensation. At the higher temperature, these liquid bridges between the particles can exert enough force to deform the particles.

Two MFFTs appear to exist: wet MFFT and dry MFFT. The normal, or wet, MFFT is that which is seen under normal circumstances — wherein a latex is applied at an ambient temperature above the polymer's T_g, and film formation follows the three stages described in Section 3.3. This wet MFFT is associated with particles deforming due to a receding air-water interface.

The higher temperature at which a previously uncoalesced latex deforms is the dry MFFT. This is associated with much smaller quantities of water between particles. The role of the water at this higher dry MFFT is not well understood. It may be that the smaller amounts are able to deform the particles because a different deformation mechanism is possible at the elevated temperature. Or, it may be that the polymer particle is softer under these circumstances. The phenomenon is interesting and may be helpful in improving models of latex film formation [21-24].

3.5 FLASH RUSTING

Nicholson defines flash rusting as "...the rapid corrosion of the substrate during drying of an aqueous coating, with the corrosion products (i.e., rust) appearing on the surface of the dried film" [25]. Flash rusting is commonly named as a possible drawback to waterborne coatings; yet, as Nicholson goes on to point out, the phenomenon is not understood and its long-term importance for the coating is unknown. Studies have been carried out to identify effective anti-flash-rust additives; however, because they are empirical in approach, the mechanisms by which any of them work — or even the necessity for them — has not been well defined.

The entire flash rusting discussion may be unnecessary. Igetoft [26] has pointed out that flash rusting requires not only water but also salt to be present. The fact that steel is wet does not necessarily mean that it will rust.

Forsgren and Persson [27] obtained results that seem to indicate that flash rusting is not a serious problem with modern waterborne coatings. They used contact-angle measurements, Fourier Transform Infared Spectroscopy (FTIR), and AFM to study changes in surface chemistry at the steel–waterborne acrylic coating interface before curing takes place. In particular, the total free-surface energy of the steel, and its electromagnetic and acid-base components, were studied before and immediately after application of the coating. The expectation was that the acidic or basic components, or both, of the steel's surface energy would increase immediately after the coatings were applied. Instead, the total surface energy of the steel decreased, and the Lewis base component dropped dramatically. The contact-angle measurements after contact with the coatings were more typical of polymers than of cold-rolled steel. Spectroscopy studies showed carboxyl and alkane groups on the surface of the steel after two minutes' exposure to the paint. Atomic force microscopy showed rounded particles of a softer material than steel distributed over the surface after a short exposure to the coatings. The authors speculated that the adhesion promoters on the polymer chain are so effective that the first particles of polymer are already attached to the steel after 20 seconds — in other words, before any deformation due to water evaporation could have occurred. The effects of this immediate bonding on immediate and long-term corrosion protection are unknown. Better knowledge of the processes taking place at the coating-metal interface immediately upon application of the coating may aid in understanding and preventing undesirable phenomena such as flash rusting.

REFERENCES

1. Hawkins, C.A., Sheppard, A.C., and Wood, T.G., *Prog. Org. Coat.*, 32, 253, 1997.
2. Padget, J.C., *J. Coat. Technol.*, 66, 89, 1994.
3. Misev, T.A, *J. Jap. Soc. Col. Mat.*, 65, 195, 1993.
4. Snuparek, J. et al., *J. Appl. Polym. Sci.*, 28, 1421, 1983.
5. Gauthier, C. et al., *ACS Symposium Series 648, Film Formation in Water-Borne Coatings*, Provder, T., Winnik, M.A., and Urban, M.W., Eds., American Chemical Society, Washington, 1996.
6. Gilicinski, A.G., and Hegedus, C.R., *Prog. Org. Coat.*, 32, 81, 1997.
7. Lin, F. and Meier, D.J. *Prog. Org. Coat.*, 29, 139, 1996.
8. Visschers, M., Laven, J., and van der Linde, R., *Prog. Org. Coat.*, 31, 311, 1997.
9. Visschers, M., Laven, J., and German, A.L., *Prog. Org. Coat.*, 30, 39, 1997.
10. Brown, G.L., *J. Polym. Sci.*, 22, 423, 1956.
11. Mason, G., *Br. Polym. J.*, 5, 101, 1973.
12. Lamprecht, J., *Colloid Polym. Sci.*, 258, 960, 1980.
13. Nicholson, J., *Waterborne Coatings: Oil and Colour Chemists' Association Monograph No. 2*, Oil and Colour Chemists' Association, London, 1985.
14. Franks, F., *Water*, Royal Society of Chemistry, London, 1983.
15. Forsgren, A. and Palmgren, S., *Effect of Application Climate on Physical Properties of Three Waterborne Paints,* Report 1997:3E, Swedish Corrosion Institute, Stockholm, 1997.
16. Kobayashi, T., *Prog. Org. Coat.*, 28, 79, 1996.
17. Joanicot, M., Granier, V., and Wong, K., *Prog. Org. Coat.*, 32, 109, 1997.
18. Tzitzinou, A. et al., *Prog. Org. Coat.*, 35, 89, 1999.
19. Wegmann, A., *Prog. Org. Coat.*, 32, 231, 1997.
20. Nysteen, S., Hempel's Marine Paints A/S (Denmark); personal communication.
21. Sperry, P.R. et al., *Langmuir*, 10, 2169, 1994.
22. Keddie, J.L. et al., *Macromolecules*, 28, 2673, 1995.
23. Snyder, B.S. et al., *Polym. Preprints*, 35, 299, 1994.
24. Heymans, D.M.C. and Daniel, M.F., *Polym. Adv. Technol.*, 6, 291, 1995.

25. Nicholson, J.W., in *Surface Coatings*, Wilson, A.D., Nicholson, J.W., and Prosser, H.J., Eds., Elsevier Applied Science Publ., Amsterdam, 1988, Chap. 1.

26. Igetoft, L., *Våtblästring som förbehandling före rostskyddsmålning — litteraturegenomgång*, Report 61132:1, Swedish Corrosion Institute, Stockholm, 1983. (In Swedish.)

27. Forsgren, A. and Persson, D., *Changes in the Surface Energy of Steel Caused by Acrylic Waterborne Paints Prior to Cure*, Report 2000:5E, Swedish Corrosion Institute, Stockholm, 2000.

4 Blast Cleaning and Other Heavy Surface Pretreatments

In broad terms, pretreatment of a metal surface is done for two reasons: to remove unwanted matter and to give the steel a rough surface profile before it is painted. "Unwanted matter" is anything on the surface to be painted except the metal itself and — in the case of repainting — tightly adhering old paint.

For new constructions, matter to be removed is mill scale and contaminants. The most common contaminants are transport oils and salts. Transport oils are beneficial (until you want to paint); salts are sent by an unkind Providence to plague us. Transport oil might be applied at the steel mill, for example, to provide a temporary protection to the I-beams for a bridge while they are being hauled on a flatbed truck from the mill to the construction site or the subassembly site. This oil-covered I-beam, unfortunately, acts as a magnet for dust, dirt, diesel soot, and road salts; anything that can be found on a highway will show up on that I-beam when it is time to paint. Even apart from the additional contaminants the oil picks up, the oil itself is a problem for the painter. It prevents the paint from adhering to the steel, in much the same way that oil or butter in a frying pan prevents food from sticking. Pretreatment of new steel before painting is fairly straightforward; washing with an alkali surfactant, rinsing with clean water, and then removing the mill scale with abrasive blasting is the most common approach.

Most maintenance painting jobs do not involve painting new constructions but rather repainting existing structures whose coatings have deteriorated. Surface preparation involves removing all loose paint and rust, so that only tightly adhering rust and paint are left. Mechanical pretreatments, such as needle-gun and wire brush, can remove loosely bound rust and dirt but do not provide either the cleanliness or the surface profile required for repainting the steel. Conventional dry abrasive blasting is the most commonly used pretreatment; however, wet abrasive blasting and hydrojet cleaning are excellent treatment methods that are also gaining industry acceptance.

Before any pretreatment is performed, the surface should be washed with an alkali surfactant and rinsed with clean water to remove oils and greases that may have accumulated. Regardless of which pretreatment is used, testing for chlorides (and indeed for all contaminants) is essential after pretreatment and before application of the new paint.

4.1 INTRODUCTION TO BLAST CLEANING

By far, the most common pretreatment for steel constructions prior to painting is blast cleaning, in which the work surface is bombarded repeatedly with small solid particles. If the individual abrasive particle transfers sufficient kinetic energy to the surface of the steel, it can remove mill scale, rust, clean steel, or old paint. The kinetic energy (E) of the abrasive particle before impact is defined by its mass (M) and velocity (V), as given in the familiar equation:

$$E = (MV^2)/2$$

Upon impact, this kinetic energy can be used to shatter or deform the abrasive particle, crack or deform old paint, or chip away rust. The behavior of the abrasive, as that of the old coating, depends in part on whether it favors plastic or elastic deformation.

In general, the amount of kinetic energy transferred, and whether it will suffice to remove rust, old paint, and so forth, depends on a combination of:

- Velocity and mass of the propelled abrasive particle
- Impact area
- Strength and hardness of the substrate being cleaned
- Strength and hardness of the abrasive particle

In the most-commonly used blasting technique — dry abrasive blasting — velocity of the blasting particles is controlled by the pressure of compressed air. It is more or less a constant for any given dry blasting equipment; the mass of the abrasive particle therefore determines its impact on the steel surface.

In wet abrasive blasting, in which water replaces compressed air as the propellant of the solid blasting media, velocity of the particles is governed by water pressure. In hydrojet blasting, the water itself is both the propellant and the abrasive (no solid abrasive is used). Both forms of wet blasting offer the possibility to vary the velocity by changing water pressure. It should be noted however that wet abrasive blasting is necessarily performed at much lower pressures and, therefore, velocities, than hydrojet blasting.

4.2 DRY ABRASIVE BLASTING

Only heavy abrasives can be used in preparing steel surfaces for painting. Lighter abrasive media, such as apricot kernels, plastic particles, glass beads or particles, and walnut shells, are unsuitable for heavy steel constructions. Because of their low densities, they cannot provide the amounts of kinetic energy that must be expended upon the steel's surface to perform useful work. In order to be commercially feasible, an abrasive should be:

- Heavy, so that it can bring significant amounts of kinetic energy to the substrate
- Hard, so that it doesn't shatter into dust or deform plastically (thus wasting the kinetic energy) upon impact

- Inexpensive
- Available in large quantities
- Nontoxic

4.2.1 METALLIC ABRASIVES

Steel is used as abrasive in two forms:

- Cast as round beads, or shot
- Crushed and tempered to the desired hardness to form angular steel grit

Scrap or low-quality steel is usually used, often with various additives to ensure consistent quality. Both shot and grit have good efficiency and low breakdown rates.

Steel shot and grit are used for the removal of mill scale, rust, and old paint. This abrasive can be manufactured to specification and offers uniform particle size and hardness. Steel grit and shot can be recycled 100 to 200 times. Because they generate very little dust, visibility during blasting is superior to that of most other abrasives.

Chilled iron shot or grit can be used for the removal of rust, mill scale, heat treatment scale, and old paint from forged, cast, and rolled steel. This abrasive breaks down gradually against steel substrates, so continual sieving to retain only the large particle sizes may be needed if a rough surface profile is desired in the cleaned surface.

4.2.2 NATURALLY OCCURRING ABRASIVES

Several naturally occurring nonmetallic abrasives are commercially available, including garnet, zircon, novaculite, flint, and the heavy mineral sands magnetite, staurolite, and olivine. However, not all of these abrasives can be used to prepare steel for maintenance coatings. For example, novaculite and flint contain high amounts of free silica, which makes them unsuitable for most blasting applications.

Garnet is a tough, angular blasting medium. It is found in rock deposits in Eastern Europe, Australia, and North America. With a hardness of 7 to 8 Mohr, it is the hardest of the naturally occurring abrasives and, with a specific gravity of 4.1, it is denser than all others in this class except zircon. It has very low particle breakdown on impact, thereby enabling the abrasive to be recycled several times. Among other advantages this confers, the amount of spent abrasive is minimized — an important consideration when blasting old lead- or cadmium-containing paints. The relatively high cost of garnet limits its use to applications where abrasive can be gathered for recycling. However, for applications where spent abrasive must be treated as hazardous waste, the initial higher cost of garnet is more than paid for by the savings in disposal of spent abrasive.

Nonsilica mineral sands, such as magnetite, staurolite, and olivine, are tough (5 to 7 Mohr) and fairly dense (2.0 to 3.0 specific gravity) but are generally of finer particle size than silica sand. These heavy mineral sands — as opposed to silica sand — do not contain free silicates, the cause of the disease silicosis. In general, the heavy mineral sands are effective for blast cleaning new steel but are not the best choice for maintenance applications [1].

Olivine ($[Mg,Fe]_2[SiO_4]$) has a somewhat lower efficiency than silica sand [2] and occasionally leaves white, chalk-like spots on the blasted surface. It leaves a profile of 2.5 mil or finer, which makes it less suitable for applications where profiling the steel surface is important.

Staurolite is a heavy mineral sand that has low dust levels and, in many cases, can be recycled three or four times. It has been reported to have good feathering and does not embed in the steel surface.

Zircon has higher specific gravity (4.5) than any other abrasive in this class and is very hard (7.5 Mohr). Other good attributes of zircon are its low degree of dusting and its lack of free silica. Its fine size, however, limits its use to specialty applications because it leaves little or no surface profile.

Novaculite is a siliceous rock that can be ground up to make an abrasive. It is the softest abrasive discussed in this class (4 Mohr) and is suitable only for specialty work because it leaves a smooth surface. Novaculite is composed mostly of free silica, so this abrasive is not recommended unless adequate precautions to protect the worker from silicosis can be taken. For the same reason, **flint**, which consists of 90% free silica, is not recommended for maintenance painting.

4.2.3 BY-PRODUCT ABRASIVES

By-product abrasives can be used to remove millscale on new constructions or rust and old paint in maintenance jobs. These abrasives are made from the residue, or *slag*, leftover from smelting metals or burning coal in power plants. Certain melting and boiler slags are glassy, homogeneous mixtures of various oxides with physical properties that make them good abrasives. However, not all industrial slags have the physical properties and nontoxicity needed for abrasives. Boiler (coal), copper, and nickel slags are suitable and dominate this class of abrasives. All three are angular in shape and have a hardness of 7 to 8 Mohr and a specific gravity of 2.7 to 3.3; this combination makes for efficient blast cleaning. In addition, none contain significant (1%) amounts of free silica.

Copper slag is a mixture of calcium ferrisilicate and iron orthosilicate. A by-product of the smelting and quenching processes in copper refining, the low material cost and good cutting ability of copper slag make it one of the most economical, expendable abrasives available. It is used in many industries, including major shipyards, oil and gas companies, steel fabricators, tank builders, pressure vessel fabricators, chemical process industries, and offshore yards. Copper slag is suitable for removing mill scale, rust, and old paint. Its efficiency is comparable to that of silica sand [2]. It has a slight tendency to imbed in mild steel [3].

Boiler slag — also called **coal slag** — is aluminum silicate. It has a high cutting efficiency and creates a rough surface profile. It too has a slight tendency to imbed in mild steel.

Nickel slag, like copper and boiler slag, is hard, sharp, efficient at cutting, and possesses a slight tendency to imbed in mild steel. Nickel slag is sometimes used in wet blasting (see Section 4.3).

TABLE 4.1
Physical Data for By-Product Abrasives

Abrasive	Degree of dusting	Reuse
Boiler slag	High	Poor
Copper slag	Low	Good
Nickel slag	High	Poor

Modified from: Good Painting Practice, Vol. 1, J.D. Keane, Ed.. Steel Structures Painting Council, Pittsburgh, PA, 1982.

4.2.3.1 Variations in Composition and Physical Properties

It should be noted that, because these abrasives are by-products of other industrial processes, their chemical composition and physical properties can vary widely. As a result, technical data reported can also vary widely for this class of abrasives. For example, Bjorgum has reported that copper slag created more blasting debris than nickel slag in trials done in conjunction with repainting of the Älvsborg bridge in Gothenburg, Sweden [4]. This does not agree with the information reported by Keane [1], which is shown in Table 4.1.

This contradiction in results almost certainly depends on differences in the chemical composition, hardness, and particle size of different sources of the same generic type of by-product abrasive.

Because of the very wide variations possible in chemical composition of these slags, a cautionary note should perhaps be introduced when labeling these abrasives as nontoxic. Depending on the source, the abrasive could contain small amounts of toxic metals. Chemical analyses of copper slag and nickel slag used for the Älvsborg bridge work have been reported by Bjorgum [4]. Eggen and Steinsmo have also analyzed the composition of various blasting media [5]. The results of both studies are compared in Table 4.2. Comparison of the lead levels in the nickel slags or of the zinc levels in the copper slags clearly indicates that the amounts of an element or compound can vary dramatically between batches and sources.

By-product abrasives are usually considered one-time abrasives, although there are indications that at least some of them may be recyclable. In the repainting of the Älvsborg bridge, Bjorgum found that, after one use, 80% of the particles were still larger than 250 μm; and concluded that the abrasive could be used between three and five times [4].

4.2.4 Manufactured Abrasives

The iron and steel abrasives discussed in Section 4.2.1 are of course man-made. In this section, however, we use the term "manufactured abrasives" to mean those produced for specific physical properties, such as toughness, hardness, and shape. The two abrasives discussed here are very heavy, extremely tough, and quite expensive. Their physical properties allow them to cut very hard metals, such as titanium

TABLE 4.2
Levels of Selected Compounds/Elements Found in By-Product Abrasives

Blasting media	Pb	Co	Cu	Cr	Ni	Zn
Copper slag [Eggen and Steinsmo]	0.24%	0.07%	0.14%	0.05%	71 ppm	5.50%
Copper slag [Bjorgum]	203 ppm	249 ppm	5.6 ppm	1.4 ppm	129 ppm	10 ppm
Nickel slag [Eggen and Steinsmo]	73 ppm	0.43%	0.28%	0.14%	0.24%	0.38%
Nickel slag [Bjorgum]	1.2 ppm	2.3 ppm	4.5 ppm	755 ppm	1.1 ppm	15.6 ppm

Sources: Bjorgum, A., *Behandling av avfall fra bläserensing, del 3. Oppsummering av utredninger vedrorende behandling av avfall fra blåserensing,* Report No. STF24 A95326, Foundation for Scientific and Industrial Research at the Norwegian Institute of Technology (SINTEF), Trondheim, 1995 (in Norwegian); Eggen, T. and Steinsmo, U., *Karakterisering av flater blast med ulike blåsemidler,* Report No. STF24 A94628, SINTEF, Trondheim, 1994 (in Norwegian).

and stainless steel, and to be recycled many times before significant particle breakdown occurs.

Manufactured abrasives are more costly than by-product slags, usually by an order of magnitude. However, the good mechanical properties of most manufactured abrasives make them particularly adaptable for recycling as many as 20 times. In closed-blasting applications where recycling is designed into the system, these abrasives are economically attractive. Another important use for them is in removing old paints containing lead, cadmium, or chromium. When spent abrasive is contaminated with these hazardous substances, the abrasive might need to be treated and disposed of as a hazardous material. If disposal costs are high, an abrasive that generates a low volume of waste — due to repeated recycling — gains in interest.

Silicon carbide, or **carborundum**, is a dense and extremely hard angular abrasive (specific gravity 3.2, 9 Mohr). It cleans extremely fast and generates a rough surface profile. This abrasive is used for cleaning very hard surfaces. Despite its name, it does not contain free silica.

Aluminium oxide is a very dense and extremely hard angular abrasive (specific gravity 4.0, 8.5 to 9 Mohr). It provides fast cutting and a good surface profile so that paint can anchor onto steel. This abrasive generates low amounts of dust and can be recycled, which is necessary because it is quite expensive. Aluminium oxide does not contain free silica.

4.3 WET ABRASIVE BLASTING AND HYDROJETTING

In dry abrasive blasting, a solid abrasive is entrained in a stream of compressed air. In **wet abrasive blasting,** water is added to the solid abrasive medium. Another approach is to keep the water but remove the abrasive; this is called **hydrojetting,** or **water jetting.** This pretreatment method depends entirely on water impacting a steel surface at a high enough speed to remove old coatings, rust, and impurities.

The presence of an abrasive medium in the dry or wet pretreatment methods results in a surface with a desirable profile. Hydrojetting, on the other hand, does not increase the surface roughness of the steel. This means that hydrojetting is not suitable for new constructions because the steel will never receive the surface roughness necessary to provide good anchoring of the paint. For repainting or maintenance painting, however, hydrojetting may be used to strip away paint, rust, and so forth and restore the original surface profile of the steel.

Paul [6] mentions that because dust generation is greatly reduced in wet blasting, this method makes feasible the use of some abrasives that would otherwise be health hazards. This should not be taken as an argument to use health-hazardous abrasives, however, because more user-friendly abrasives are available in the market.

4.3.1 TERMINOLOGY

The terminology of wet blasting is confusing, to say the least. The following useful definitions are found in the *Industrial Lead Paint Removal Handbook* [7]:

- *Wet abrasive blast cleaning*: Compressed air propels abrasive against the surface. Water is injected into the abrasive stream either before or after the abrasive exits the nozzle. The abrasive, paint debris, and water are collected for disposal.
- *High-pressure water jetting*: Pressurized water (up to 20,000 psi) is directed against the surface to remove the paint. Abrasives are not used.
- *High-pressure water jetting with abrasive injection*: Pressurized water (up to 20,000 psi) is directed against the surface to be cleaned. Abrasive is metered into the water stream to facilitate the removal of rust and mill scale and to improve the efficiency of paint removal. Disposable abrasives are used.
- *Ultra-high-pressure water jetting*: Pressurized water (20,000–40,000 psi; can be higher) is directed against the surface to remove the paint. Abrasives are not used.
- *Ultra-high-pressure water jetting with abrasive injection*: Pressurized water (20,000–40,000; can be greater) is directed against the surface to be cleaned. Abrasive is metered into the water stream to facilitate the removal of rust and mill scale and to improve the efficiency of paint removal. Disposable abrasives are used.

4.3.2 INHIBITORS

An important question in the area of wet blasting is does the flash rust, which can appear on wet-blasted surfaces, have any long-term consequences for the service life of the subsequent painting? A possible preventative for flash rust is adding a corrosion inhibitor to the water.

The literature on rust inhibitors is mixed. Some sources view them as quite effective against corrosion, although they also have some undesirable effects when properly used. Others, however, view rust inhibitors as a definite disadvantage. Which chemicals are suitable inhibitors is also an area of much discussion.

Sharp [8] lists nitrites, amines, and phosphates as common materials used to make inhibitors. He notes problems with each class:

- If run-off water has a low pH (5.5 or less), nitrite-based inhibitors can cause the residue to form a weak but toxic nitrous oxide, which is a safety concern for workers.
- Amine-based inhibitors can lose some of their inhibitive qualities in low-pH environments.
- When using ultra-high pressure, high temperatures at the nozzle (greater than 140°F [60°C]) can cause some phosphate-based inhibitors to revert to phosphoric acid, resulting in a contaminant build-up.

In the 1966 edition of the manual *Good Painting Practice,* the Steel Structures Painting Council recommended an inhibitor made of diammonium phosphate and sodium nitrite [9]. Other possibilities include chromic acid, sodium chromate, sodium dichromate, and calcium dichromate. The 1982 edition of this manual does not make detailed recommendations of specific inhibitor systems [1].

Van Oeteren [10] lists the following possible inhibitors:

- Sodium nitrite combined with sodium carbonate or sodium phosphate
- Sodium benzoate
- Phosphate, alkali (sodium phosphate or hexametasodium phosphate)
- Phosphoric acid combinations
- Water glass

He also makes the important point that hygroscopic salts under a coating lead to blistering and that, therefore, only inhibitors that do not form hygroscopic salts should be used for wet blasting.

McKelvie [11] does not recommend inhibitors for two reasons. First, flash rusting is useful in that it is an indication that salts are still present on the steel surface; and second, he also points out that inhibitor residue on the steel surface can cause blistering.

The entire debate over inhibitor use may be unnecessary. Igetoft [12] points out that the amount of flash rusting of a steel surface depends not only on the presence of water but also very much on the amount of salt present. The implications of his point seem to be this: if wet blasting does a sufficiently good job of removing contaminants from the surface, the fact that the steel is wet afterward does not necessarily mean that it will rust.

4.3.3 Advantages and Disadvantages of Wet Blasting

Wet blasting has both advantages and disadvantages. Some of the advantages are:

- More salt is removed with wet blasting (see 4.3.4).
- Little or no dust forms. This is advantageous both for protection of personnel and nearby equipment, and because the blasted surface will not be contaminated by dust.

TABLE 4.3
Chloride Levels Left after Various Pretreatments

Pretreatment Method	Mean Chloride Concentration (mg/m2)		% Chloride Removal
	Before Pretreatment	After Pretreatment	
Hand wirebrush to grade St 3	157.0	152.0	3
Needlegun to grade St 3	116.9	113.5	3
Ultra-high-pressure (UHP) waterjet to grade DW 2	270.6	17.8	93
UHP waterjet to grade DW 3	241.9	15.7	94
Dry grit-blasting to Sa 2 1/2	211.6	33.0	84

Source: Allen, B., *Prot. Coat. Eur.,* 2, 38, 1997.

- Precision blasting, or blasting a certain area without affecting nearby areas of the surface, is possible.
- Other work can be done in the vicinity of wet blasting.

Among the disadvantages reported are:

- Equipment costs are high.
- Workers have limited vision in and general difficulties in accessing enclosed spaces.
- Clean up is more difficult.
- Drying is necessary before painting.
- Flash rusting can occur (although this is debatable [see Section 4.3.1])

4.3.4 CHLORIDE REMOVAL

As part of a project testing surface preparation methods for old, rusted steel, Allen [13] examined salt contamination levels before and after treating the panels. Hydrojetting was found to be the most effective method for removing salt, as can be seen in Table 4.3.

The Swedish Corrosion Institute found similar results in a study on pretreating rusted steel [14]. In this study, panels of hot-rolled steel, from which the mill scale had been removed using dry abrasive blasting, were sprayed daily with 3% sodium chloride solution for five months, until the surface was covered with a thick, tightly adhering layer of rust. Panels were then subjected to various pretreatments to remove as much rust as possible and were later tested for chlorides with the Bresle test. Results are given in Table 4.4.

4.3.5 WATER CONTAINMENT

Containment of the water used for pressure washing is an important concern. If used to remove lead-based paint, the water may contain suspended lead particles and needs to be tested for leachable lead using the toxicity characteristic leaching procedure

TABLE 4.4
Chloride Levels after Various Pretreatments

Pretreatment Method	Average Chloride Level (mg/m²)	% Chloride Removal
No pretreatment	349	--
Wirebrush to grade SB2	214	39
Needlegun to grade SB2	263	25
UHP hydrojet, 2500 bar, no inhibitor	10	97
Wet blasting with aluminium silicate abrasive, 300 bar, no inhibitor	16	95
Dry grit-blasting to Sa 2 1/2 (copper slag)	56	84

Source: Forsgren, A. and Appelgren, C., Comparison of Chloride Levels Remaining on the Steel Surface after Various Pretreatments, *Proc. Pro. Coat. Eur. 2000*, Technology Publishing Company, 2000, 271.

(see Chapter 5) prior to discharge. Similarly, testing before discharge is needed when using wet blasting or hydrojetting to remove cadmium- or chromium-pigmented coatings. If small quantities of water are used, it may be acceptable to pond the water until the testing can be conducted [13].

4.4 UNCONVENTIONAL BLASTING METHODS

Dry abrasive blasting will not disappear in the foreseeable future. However, other blasting techniques are currently of interest. Some are briefly described in this section: dry blasting with solid carbon dioxide, dry blasting with an ice abrasive, and wet blasting with soda as an abrasive.

4.4.1 CARBON DIOXIDE

Rice-sized pellets of carbon dioxide (dry ice) are flung with compressed air against a surface to be cleaned. The abrasive sublimes from solid to gas phase, leaving only paint debris for disposal. This method reportedly produces lower amounts of dust, and thus containment requirements are reduced. Workers are still exposed to any heavy metals that exist in the paint and must be protected against them.

Disadvantages of this method are its high equipment costs and slow removal of paint. In addition, large amounts of liquid carbon dioxide (i.e., a tanker truck) are needed. Special equipment is needed both for production of the solid carbon dioxide grains and for blasting. Although carbon dioxide is a greenhouse gas, the total amount of carbon dioxide emissions need not increase if a proper source is used. For example, if carbon dioxide produced by a fossil-fuel-burning power station is used, the amount of carbon dioxide emitted to the atmosphere does not increase.

This method can be used to remove paint but is ineffective on mill scale and heavy rust. If the original surface was blast cleaned, the profile is often restored after dry-ice blasting. As Trimber [7] sums up, "Carbon dioxide blast cleaning is an excellent concept and may represent trends in removal methods of the future."

4.4.2 ICE PARTICLES

Ice is used for cleaning delicate or fragile substrates, for example, painted plastic composites used in aircrafts. Ice particles are nonabrasive; the paint is removed when the ice causes fractures in the coating upon impact. The ice particles' kinetic energy is transferred to the coating layer and causes conical cracks, more or less perpendicular to the substrate; then lateral and radial cracks develop. When the crack network has developed sufficiently, a bit of coating flakes off. The ice particles then begin cracking the newly exposed paint that was underneath the paint that flaked off. Water from the melted ice rinses the surface free from paint flakes.

Foster and Visaisouk [15] have reported that this technique is good for removing contaminants from crevices in the blasted surfaces. Other advantages are [15]:

* Ice is nonabrasive and masking of delicate surfaces is frequently unnecessary.
* No dust results from breakdown of the blasting media.
* Ice melts to water, which is easily separated from paint debris.
* Ice can be produced on-site if water and electricity are available.
* Escaping ice particles cause much less damage to nearby equipment than abrasive media.

Ice-particle blasting has been tested for cleaning of painted compressor and turbine blades on an aircraft motor. The technique successfully removed combustion and corrosion products. The method has also been tested on removal of hydraulic fluid from aircraft paint (polyurethane topcoat) and removal of polyurethane topcoat and epoxy primer from an epoxy graphite composite.

4.4.3 SODA

Compressed air or high-pressure water is used to propel abrasive particles of sodium bicarbonate against a surface to be cleaned. Sodium bicarbonate is water-soluble; paint chips and lead can be separated from the water and dissolved sodium bicarbonate, thereby reducing the volume of hazardous waste.

The water used with sodium bicarbonate significantly reduces dust. The debris is comprised of paint chips, although it may also be necessary to dispose of the water and dissolved sodium bicarbonate as a hazardous waste unless the lead can be completely removed. The need to capture water can create some difficulties for containment design.

This technique is effective at removing paint but cannot remove mill scale and heavy corrosion. In addition, the quality of the cleaning may not be suitable for some paint systems, unless the surface had been previously blast-cleaned. If bare steel is exposed, inhibitors may be necessary to prevent flash rusting.

Most painting contractors are not familiar with this method but, because of similarities to wet abrasive blasting and hydrojetting, they can easily adjust. Because the water mitigates the dust, exposure to airborne lead emissions is significantly reduced but not eliminated; ingestion hazards still exist [15].

4.5 TESTING FOR CONTAMINANTS AFTER BLASTING

Whichever pretreatment method is used, it is necessary before painting to check that the metal surface is free from salts, oils, and dirt.

4.5.1 SOLUBLE SALTS

No matter how good a new coating is, applying it over a chloride-contaminated surface is begging for trouble. Chloride contamination can occur from a remarkable number of sources, including road salts if the construction is anywhere near a road or driveway that is salted in the winter. Another major source for constructions in coastal areas is the wind; the tangy, refreshing feel of a sea breeze means repainting often if the construction is not sheltered from the wind. Even the hands of workers preparing the steel for painting contain enough salt to cause blistering after the coating is applied.

Rust in old steel can also be a major source of chlorides. The chlorides that originally caused the rust are caught up in the rust matrix; by their very nature, in fact, chlorides exist at the bottom of corrosion pits — the hardest place to reach when cleaning [16,17].

The ideal test of soluble salts is an apparatus that could be used for nondestructing sampling:

- On-site rather than in the lab
- On all sorts of surfaces (rough, smooth; curved, flat)
- Quickly, because time is money
- Easily, with results that are not open to misinterpretation
- Reliably
- Inexpensively

Such an instrument does not exist. Although no single method combines all of these attributes, some do make a very good attempt. All rely upon wetting the surface to leach out chlorides and other salts and then measuring the conductivity of the liquid, or its chloride content, afterward. Perhaps the two most-commonly used methods are the Bresle patch and the wetted-filter-paper approach from Elcometer.

The Bresle method is described in the international standard ISO 8502-6. A patch with adhesive around the edges is glued onto the test surface. This patch has a known contact area, usually 1250 mm². A known volume of deionized water is injected into the cell. After the water has been in contact with the steel for 10 minutes, it is withdrawn and analyzed for chlorides. There are several choices for analyzing chloride content: titrating on-site with a known test solution; using a conductivity meter; or where facilities permit, using a more sophisticated chloride analyzer. Conductivity meters cannot distinguish between chemical species. If used on heavily rusted steel, the meter cannot distinguish how much of the conductivity is due to chlorides and how much is due simply to ferrous ions in the test water.

The Bresle method is robust; it can be used on very uneven or curved surfaces. The technique is easy to perform, and the equipment inexpensive. Its major drawback is the time it requires; 10 minutes for a test is commonly believed to be too long, and there is a strong desire for something as robust and reliable — but faster.

The filter-paper technique is much faster. A piece of filter paper is placed on the surface to be tested, and deionized water is squirted on it until it is saturated. The wet paper is then placed on an instrument (such as the SCM-400 from Elcometer) that measures its resistivity. As in the conductivity measurements discussed above, when this is used for repainting applications, it is not certain how much of the resistivity of the paper is due to chlorides and how much is due simply to rust in the test water. In all, the technique is reliable and simple to implement, although initial equipment costs are rather high.

Neither technique measures all the chlorides present in steel. The Bresle technique is estimated to have around 50% leaching efficiency; the filter paper technique is somewhat higher. One could argue, however, that absolute values are of very limited use; if chlorides are present in any quantity, they will cause problems for the paint. It does not perhaps matter at all that a measurement technique reports 200 mg/m^2, when the correct number was 300 mg/m^2. Both are far too high. This, indeed, is a weakness in the field of pretreatment quality control; it is not known how much chlorides is too much. Although there is some consensus that the acceptable amount is very low, there is no consensus on what the cut-off value is [18–20].

4.5.2 HYDROCARBONS

Like salts, hydrocarbons in the form of oils and grease also come from a variety of sources: diesel fumes, either from passing traffic or stationary equipment motors; lubricating oils from compressors and power tools; grease or oil in contaminated blasting abrasive; oil on operators' hands; and so on. As mentioned above, the presence of oils and grease on the surface to be painted prevents good adhesion.

Testing for hydrocarbons is more complex than is testing for salts for two reasons. First, hydrocarbons are organic, and organic chemistry in general is much more complex than the inorganic chemistry of salts. A simple indicator kit of reagents is quite tricky to develop when organic chemistry is involved. Second, a vast range of hydrocarbons can contaminate a surface, and a test that checks for just a few of them would be fairly useless. What is needed, then, is a test simple enough to be done in the field and powerful enough to detect a broad range of hydrocarbons.

Ever game, scientists have developed a number of approaches for testing for hydrocarbons. One approach is ultraviolet (UV) light, or black lights. Most hydrocarbons show up as an unappetizing yellow or green under a UV lamp. This only works, of course, in the dark and, therefore, testing is done under a black hood, rather like turn-of-the-century photography. Drawbacks are that lint and possibly dust show up as hydrocarbon contaminations. In addition, some oils are not detected by black lights [21]. In general, however, this method is easier to use than other methods.

Other methods that are currently being developed for detecting oils include [22]:

- **Iodine with the Bresle patch.** Sampling is performed according to the Bresle method (blister patch and hypodermic), but with different leaching liquids. The test surface is first prepared with an aqueous solution of iodine and then washed with distilled water. Extraction of the dissolved iodine in oil on the surface is thereafter made by the aid of a potassium iodide solution. After extraction of the initially absorbed iodine from the contaminated surface, starch is added to the potassium iodide solution. Assessment of the amount of iodine extracted from the surface is then determined from the degree of blue coloring of the solution. Because the extracted amount of iodine is a measure of the amount of oil residues on the surface, the concentration of the oil on the surface can be determined.
- **Fingerprint tracing method.** Solid sorbent of aluminum oxide powder is spread over the test surface. After heat treatment, the excess of sorbent not strongly attached to the contaminated surface is removed. The amount of attached sorbent is thereafter scraped off the surface and weighed. This amount of sorbent is a measure of the amount of oil or grease residues on the surface.
- **Sulfuric acid method.** For extraction of oil and grease residues from the surface, a solid sorbent aluminum oxide is used here, too. However, concentrated sulfuric acid is added to the aluminum oxide powder that is scraped off from the contaminated surface. The sulfuric acid solution with the extracted oil and grease residues is then heated. From the coloring of the solution, which varies from colorless to dark brown, the amount of oil and grease residues can be determined.

4.5.3 Dust

Dust comes from the abrasive used in blasting. All blasting abrasives break down to some extent when they impact the surface being cleaned. Larger particles fall to the floor, but the smallest particles form a dust too fine to be seen. These particles are held on the surface by static electricity and, if not removed before painting, prevent the coating from obtaining good adhesion to the substrate.

Examining the surface for dust is straightforward: wipe the surface with a clean cloth. If the cloth comes away dirty, then the surface is too contaminated to be painted. Another method is to apply tape to the surface to be coated. If the tape, when pulled off, has an excessive amount of fine particles attached to the sticky side, then the surface is contaminated by dust. It is a judgment call to say whether a surface is too contaminated because, for all practical purposes, it is impossible to remove all dust after conventional abrasive blasting.

Testing for dust should be done at every step of the paint process because contamination can easily occur after a coating layer has been applied, causing the paint to become tack-free. This would prevent good adhesion of the next coating layer.

Dust can be removed by vacuuming or by blowing the surface down with air. The compressed air used must be clean — compressors are a major source of oil contamination. To check that the compressed air line does not contain oil, hold a clean piece of white paper in front of the air stream. If the paper becomes dirtied with oil (or water, or indeed anything else), the air is not clean enough to blow down the surface before painting. Clean the traps and separators and retest until the air is clean and free from water [21].

4.6 DANGEROUS DUST: SILICOSIS AND FREE SILICA

Dry abrasive blasting with silica sand is banned or restricted in many countries because of its link to the disease silicosis, which is caused by breathing excessive quantities of extremely fine particles of silica dust over a long period. This section discusses:

- What silicosis is
- What forms of silicon cause silicosis
- Low-free-silica abrasive options
- Hygienic measures to prevent silicosis

4.6.1 WHAT IS SILICOSIS?

Silicosis is a fibronodular lung disease caused by inhaling dust containing crystalline silica. When particles of crystalline silica less than 1 μm are inhaled, they can penetrate deeply into the lungs, through the bronchioles and down to the alveoli. When deposited on the alveoli, silica causes production of radicals that damage the cell membrane. The alveoli respond with inflammation, which damages more cells. Fibrotic nodules and scarring develop around the silica particles. As the amount of damage becomes significant, the volume of air that can flow through the lungs decreases and, eventually, respiratory failure develops. Epidemiologic studies have established that patients with silicosis are also more vulnerable to tuberculosis; the combination of diseases is called silicotuberculosis and has an increased mortality over silicosis [23–26].

Silicosis has been recognized since 1705, when it was remarked among stone-cutters. It has long been recognized as a grave hazard in certain occupations, for example, mining and tunnel-boring. The worst known epidemic of silicosis was in the drilling of the Gauley Bridge Tunnel in West Virginia in the 1930s. During the construction, an estimated 2,000 men were involved in drilling through the rock. Four hundred died of silicosis; of the remaining 1,600, almost all developed the disease.

Silicosis is of great concern to abrasive blasters, because the silica breaks down upon impact with the surface being cleaned. The freshly fractured surfaces of silica appear to produce more severe reactions in the lungs than does silica that is not newly fractured [27], probably because the newly split surface of silica is more chemically reactive.

4.6.2 What Forms of Silica Cause Silicosis?

Not all forms of silica cause silicosis. Silicates are not implicated in the disease, and neither is the element silicon (Si), commonly distributed in the earth's crust and made famous by the semiconductor industry.

Silicates (-SiO$_4$) are a combination of silicon, oxygen, and a metal such as aluminum, magnesium, or lead. Examples are mica, talc, Portland cement, asbestos, and fiberglass.

Silica is silicon and oxygen (SiO$_2$). It is a chemically inert solid that can be either amorphous or crystalline. **Crystalline silica, also called "free silica," is the form that causes silicosis.** Free silica has several crystalline structures, the most common of which (for industrial purposes) are quartz, tridymite, and cristobalite. Crystalline silica is found in many minerals, such as granite and feldspar, and is a principal component of quartz sand. Although it is chemically inert, it can be a hazardous material and should always be treated with respect.

4.6.3 What is a Low-Free-Silica Abrasive?

A low-free-silica abrasive is one that contains less than 1% free (crystalline) silica. The following are examples of low-free-silica abrasives used in heavy industry:

- Steel or chilled iron, in grit or shot form
- Copper slag
- Boiler slag (aluminium silicate)
- Nickel slag
- Garnet
- Silicon carbide (carborundum)
- Aluminium oxide

4.6.4 What Hygienic Measures Can Be Taken to Prevent Silicosis?

The best way to prevent silicosis among abrasive blasters is to use a low-free-silica abrasive. Good alternatives to quartz sand are available (see Section 4.2). In many countries where dry blasting with quartz sand is forbidden, these alternatives have proven themselves reliable and economical for many decades.

It is possible to reduce the risks associated with dry abrasive blasting with silica. Efforts needed to do so can be divided into four groups:

- Less-toxic abrasive blasting materials
- Engineering controls (such as ventilation) and work practices
- Proper and adequate respiratory protection for workers
- Medical surveillance programs

The National Institute for Occupational Safety and Health (NIOSH) recommends the following measures to reduce crystalline silica exposures in the workplace and prevent silicosis [28]:

- Prohibit silica sand (and other substances containing more than 1% crystalline silica) as an abrasive blasting material and substitute less hazardous materials.
- Conduct air monitoring to measure worker exposures.

- Use containment methods, such as blast-cleaning machines and cabinets, to control the hazard and protect adjacent workers from exposure.
- Practice good personal hygiene to avoid unnecessary exposure to silica dust.
- Wear washable or disposable protective clothes at the worksite; shower and change into clean clothes before leaving the worksite to prevent contamination of cars, homes, and other work areas.
- Use respiratory protection when source controls cannot keep silica exposures below the NIOSH Recommended Exposure Limit.
- Provide periodic medical examinations for all workers who may be exposed to crystalline silica.
- Post signs to warn workers about the hazard and inform them about required protective equipment.
- Provide workers with training that includes information about health effects, work practices, and protective equipment for crystalline silica.
- Report all cases of silicosis to state health departments and to the Occupational Safety and Health Administration (OSHA) or the Mine Safety and Health Administration.

For more information, the interested reader is encouraged to obtain the free document, *Request for Assistance in Preventing Silicosis and Deaths from Sandblasting*, DHHS (NIOSH) Publication No 92-102, which is available at www.osha.gov or by contacting NIOSH at the following address:

> Information Dissemination Section
> Division of Standards Development
> and Technology Transfer
> NIOSH
> 4676 Columbia Parkway
> Cincinnati, Ohio 45226 USA

REFERENCES

1. *Good Painting Practice,* Vol. 1, Keane, J.D., Ed., Steel Structures Painting Council, Pittsburgh, PA, 1982.
2. *Handbok i rostskyddsmålning av allmänna stålkonstruktioner.* Bulletin Nr. 85, 2nd ed., Swedish Corrosion Institute, Stockholm, 1985. (In Swedish.)
3. *Evaluation of copper slag blast media for railcar maintenance*, NASA-CR-183744, N90-13681, National Aeronautics and Space Administration, George C. Marshall Space Flight Center, AL, 1989.
4. Bjorgum, A., *Behandling av avfall fra bläserensing, del 3. Oppsummering av utredninger vedrorende behandling av avfall fra blåserensing.* Report No. STF24 A95326, Foundation for Scientific and Industrial Research at the Norwegian Institute of Technology, Trondheim, 1995. (In Norwegian.)
5. Eggen, T. and Steinsmo, U., *Karakterisering av flater blast med ulike blåsemidler*, Report No. STF24 A94628, Foundation for Scientific and Industrial Research at the Norwegian Institute of Technology, Trondheim, 1994. (In Norwegian.)

6. Paul, S., *Surface Coatings Science and Technology.* 2nd ed. John Wiley & Sons, Chichester, England, 1996.

7. Trimber, K.A., *Industrial Lead Paint Removal Handbook*, SSPC 93-02, Steel Structures Painting Council, Pittsburgh, PA, 1993, Chapters 1-9.

8. Sharp, T., *J. Prot. Coat. Linings*, 13, 133, 1996.

9. *Good Painting Practice*, Vol. 1., 1st ed. Steel Structures Painting Council, Pittsburgh, PA, 1966.

10. van Oeteren, K.A., *Korrosionsschutz durch Beschichtungsstoffe, part 1.* Carl-Hanser Verlag, Munich. 1980.

11. McKelvie, A.N., Planning and control of corrosion protection in shipbuilding, in *Proceedings 6th International Congress on Metallic Corrosion,* Sydney, 1975, Paper 8-7.

12. Igetoft, L. "Våtblästring som förbehandling före rostskyddsmålning - litteraturegenomgång," Report No. 61132:1, Swedish Corrosion Institute, Stockholm, 1983. (In Swedish.)

13. Allen, B., *Prot. Coat. Eur.*, 2, 38, 1997.

14. Forsgren, A., and Appelgren, C., Comparison of chloride levels remaining on the steel surface after various pretreatments, *Proc. Pro. Coat. Evr. 2000*, Technology Publishing Company, Pittsburgh, 2000, 271.

15. Foster, T. and Visaisouk, S., Paint removal and surface cleaning using ice particles, *Proc. AGARD SMP Lecture Series on "Environmentally Safe and Effective Processes for Paint Removal,"* NTIS 95-32171, Washington, 1995.

16. Mayne, J.E.O., *J. Appl. Chem.*, 9, 673, 1959.

17. Appleman, B.R., *J. Prot. Coat. Linings*, 4, 68, 1987.

18. Igetoft, L., *Proc. 2nd World Congress: Coatings Systems Bridges*, University of Missouri, Rolla, MO, 1982.

19. West, J., presentation, *UK Corrosion '85*, Harrogate, 4-6 November 1985. Cited in Thomas, N.L., *Proc. PRA Symposium "Coatings for Difficult Surfaces,"* Harrogate, 1990, Paper Nr. 10.

20. Morcillo, M., et al. *J. Protective Coatings and Linings*, 4, 38, 1987.

21. Swain, J.B., *J. Prot. Coat. Linings*, 4, 51, 1987.

22. Carlsson, B., Report #SP AR 1997:19, Swedish National Testing and Research Institute. SP, Borås, Sweden. 1997.

23. Myers, C.E., Hayden, C., and Morgan, J., Clinical experience with silicotuberculosis, *Penn. Med.*, 60–62, March 1973.

24. Sherson, D. and Lander, F., *J. Occup. Med.*, 32, 111, 1990.

25. Bailey, W.C. et al., *Am. Rev. Respir. Dis.*, 110, 115, 1974.

26. Silicosis and Silicate Disease Committee, Diseases associated with exposure to silica and nonfibrous silicate minerals, *Arch. Pathol. Lab. Med.*, 112, 673, 1988.

27. Vallyathan, V. et al., *Am. Rev. Respir. Dis.*, 138, 1213, 1988.

28. National Institute of Occupational Safety and Health, NIOSH Alert: Request for assistance in preventing silicosis and deaths from sandblasting, Publication (NIOSH) 92-102, U.S. Department of Health and Human Services, Cincinnati, OH, 1992.

5 Abrasive Blasting and Heavy-Metal Contamination

In the previous chapter, mention was made of the need to minimize spent abrasive when blasting old coatings containing lead pigments. This chapter covers some commonly used techniques to detect lead, chromium, and cadmium in spent abrasive and methods for disposing of abrasive contaminated with lead-based paint (LBP) chip or dust. Lead receives the most attention, both in this chapter and in the technical literature. This is not surprising because the amount of lead in coatings still in service dwarfs that of cadmium, barium, or chromium.

The growing body of literature on the treatment of lead-contaminated abrasive seldom distinguishes between the various forms of lead found in old coatings, although toxicology literature is careful to do so. Red lead (Pb_3O_4), for example, is the most common lead pigment in old primers, and white lead ($PbCO_3 \bullet Pb[OH]_2$) is more commonly found in old topcoats. It is unknown whether or not these two lead pigments will leach out at the same rate once they are in landfills. It is also unknown whether they will respond to stabilization or immobilization treatments in a similar manner. A great deal of research remains to be done in this area.

5.1 DETECTING CONTAMINATION

There are really two questions involved in detecting the presence of lead or other heavy metals:

1. Does the old paint being removed contain heavy metals?
2. Will the lead leach out from a landfill?

The amount of a metal present in paint is not necessarily the amount that will leach out when the contaminated blasting media and paint has been placed in a landfill [1-3]. The rate at which a toxic metal leaches out depends on many factors. At first, leaching comes from the surface of the paint particles. The initial rate, therefore, depends most on the particle size of the pulverized paint. This in turn depends on the condition of the paint to be removed, the type of abrasive used, and the blasting process used [4]. Eventually, as the polymeric backbone of the paint breaks down in a landfill, leaching comes from the bulk of the disintegrating paint particles. The rate at which this happens depends more on the type of resin used in formulating the paint and its chemistry in the environment of the landfill.

5.1.1 CHEMICAL ANALYSIS TECHNIQUES FOR HEAVY METALS

Several techniques are available for determining whether or not toxic metals, such as lead and chromium, exist in paint. Some well-established methods, particularly for lead, are atomic absorption (AA) and inductively coupled plasma atomic emission spectroscopy (ICP or ICP-AES). Energy-dispersive x-ray in conjunction with scanning electron microscopy (EDX-SEM) is a somewhat newer technique.

In the AA and ICP-AES methods, paint chips are dissolved by acid digestion. The amount of heavy metals in the liquid is then measured by AA or ICP-AES analysis. The amount of lead, cadmium, and other heavy metals can be calculated — with a high degree of accuracy — as a total weight percent of the paint. A very powerful advantage of this technique is that it can be used to analyze an entire coating system, without the need to separate and study each layer. Also, because the entire coating layer is dissolved in the acid solution, this method is unaffected by stratification of heavy metals throughout the layer. That is, there is no need to worry about whether the lead is contained mostly in the bulk of the layer, at the coating-metal interface, or at the topmost surface.

EDX-SEM can be used to analyze paint chips quickly. The technique is only semiquantitative: it is very capable of identifying whether the metals of interest are present but is ineffective at determining precisely how much is present. Elements from boron and heavier can be detected. EDX-SEM examines only the surface of a paint chip, to a depth of approximately 5 μm. This is a drawback because the surface usually consists of only binder. It may be possible to use very fine sandpaper to remove the top layer of polymer from the paint; however, this would have to be done very carefully so as not to sand away the entire paint layer. Of course, if the coating has aged a great deal and is chalking, then the topmost polymer layer is already gone. Therefore, analyzing cross-sections of paint chips is unnecessary in many cases, particularly for systems with two or more coats. Because coatings are not homogeneous, several measurements should be taken.

5.1.2 TOXICITY CHARACTERISTIC LEACHING PROCEDURE

Toxicity characteristic leaching procedure (TCLP) is the method mandated by the U.S, Environmental Protection Agency (EPA) for determining how much toxic material is likely to leach out of solid wastes. A short description of the TCLP method is provided here. For an exact description of the process, the reader should study Method 1311 in EPA Publication SW-846 [5].

In TCLP, a 100g sample of debris is crushed until the entire sample passes through a 9.5 mm standard sieve. Then 5 g of the crushed sample are taken to determine which extraction fluid will be used. Deionized water is added to the 5g sample to make 100 ml of solution. The liquid is stirred for 5 minutes. After that time, the pH is measured. The pH determines which extraction fluid will be used in subsequent steps, as shown in Table 5.1. The procedure for making the extraction fluids is shown in Table 5.2. The debris sample and the extraction fluid are combined and placed in a special holder. The holder is rotated at 30 ± 2 RPM for 18 ± 2 hours. The temperature is maintained at 23 ± 2°C during this time.

TABLE 5.1
pH Measurement to Determine TCLP Extraction Fluid

If the first pH measurement is:	...then
< 5.0	Extraction fluid #1 is used.
> 5.0	Acid is added. The solution is heated and then allowed to cool. Once the solution cools, pH is measured again (see below).

If the second pH measurement is:	...then
< 5.0	Extraction Fluid #1 is used.
> 5.0	Extraction Fluid #2 is used.

The liquid is then filtered and analyzed. Analysis for lead and heavy metals is done with AA or ICP-AES.

TCLP is an established procedure, but more knowledge about the chemistry involved in spent abrasive disposal is still needed. Drozdz and colleagues have reported that, in the TCLP procedure, the concentrations of lead in basic lead silico chromate are suppressed below the detection limit if zinc potassium chromate is also present. The measured levels of chromium are also suppressed, although not below the detection limit. They attribute this reduction to a reaction between the two pigments that produces a less-soluble compound or complex of lead [6].

5.2 MINIMIZING THE VOLUME OF HAZARDOUS DEBRIS

In chapter 4, we mentioned that choosing an abrasive that could be recycled several times could minimize the amount of spent abrasive. The methods described here attempt to further reduce the amount that must be treated as hazardous debris by

TABLE 5.2
Extraction Fluids for TCLP Procedure

	Extraction Fluid #1	Extraction Fluid #2
Step 1	5.7 ml glacial acetic acid is added to 500 ml water.	5.7 ml glacial acetic acid is added to water (water volume < 990 ml).
Step 2	64.3 ml sodium hydroxide is added.	Water is added until the volume is 1 L.
Step 3	Water is added until the volume is 1 L.	
Final pH	4.93 ± 0.05	2.88 ± 0.05

Note: Water used is ASTM D-1193 Type II.

separating out heavy metals from the innocuous abrasive and paint binder. The approaches used are:

- Physical separation
- Burning off the innocuous parts
- Acid extraction and then precipitation of the metals

At the present time, none of these methods is feasible for the quantities or types of heavy abrasives used in maintenance coatings. They are described here for those wanting a general orientation in the area of lead-contaminated blasting debris.

5.2.1 PHYSICAL SEPARATION

Methods involving physical separation depend on a difference between the physical properties (size, electromagnetics) of the abrasive and those of the paint debris. Sieving requires the abrasive particles to be different in size and electrostatic separation requires the particles to have a different response to an electric field.

5.2.1.1 Sieving

Tapscott et al. [7] and Jermyn and Wichner [8] have investigated the possibility of separating paint particles from a plastic abrasive by sieving. The plastic abrasive media presumably has vastly different mechanical properties than those of the old paint and, upon impact, is not pulverized in the same way as the coating to be removed.

The boundary used in these studies was 250 microns; material smaller than this was assumed to be hazardous waste (paint dust contaminated with heavy metals). The theory was fine, but the actual execution did not work so well. Photomicrographs showed that many extremely small particles, which the authors believe to be old paint, adhered to large plastic abrasive particles. In this case, sieving failed due to adhesive forces between the small paint particles and the larger abrasive media particles.

A general problem with this technique is the comparative size of the hazardous and nonhazardous particulate. Depending on the abrasive used and the condition of the paint, they may break down into a similar range of particle sizes. In such cases, screening or sieving techniques cannot separate the waste into hazardous and non-hazardous components.

5.2.1.2 Electrostatic Separation

Tapscott et al. [7] have also examined electrostatic separation of spent abrasive. In this process, spent plastic abrasive is injected into a high-voltage, direct-current electric field. Material separation depends on the attraction of the particles for the electric field. In theory, metal contaminants can be separated from nonmetal blasting debris. In practice, Tapscott and colleagues reported, the process sometimes produced fractions with heavier metal concentrations, but the separation was insufficient. Neither fraction could be treated as nonhazardous waste. In general, the results were erratic.

5.2.2 Low-Temperature Ashing (Oxidizable Abrasive Only)

Low-temperature ashing (LTA) can be used on oxidizable blasting debris — for example, plastic abrasive — to achieve a high degree of volume reduction in the waste. Trials performed with this technique on plastic abrasive resulted in a 95% reduction in the volume of solid waste. The ash remaining after oxidation must be disposed of as hazardous waste, but the volume is dramatically reduced [9].

LTA involves subjecting the spent abrasive to mild oxidation conditions at moderately elevated temperatures. The process is relatively robust: it does not depend on the mechanical properties of the waste, such as particle size, or on the pigments found in it. It is suitable for abrasives that decompose — with significant solids volume reduction — when subjected to temperatures of 500 to 600 C. Candidate abrasives include plastic media, walnut shells, and wheat starch.

The low temperature range used in LTA is thought to be more likely to completely contain hazardous components in the solid ash than is incineration at high temperatures. This belief may be unrealistic, however, given that the combustion products of paint debris mixed with plastic or agricultural abrasives are likely to be very complex mixtures [8, 9]. Studies of the mixtures generated by LTA of ground walnut shell abrasive identified at least 35 volatile organic compounds (VOCs), including propanol, methyl acetate, several methoxyphenols and other phenols, and a number of benzaldehyde and benzene compounds. In the same studies, low-temperature ashing of an acrylic abrasive generated VOCs, including alkanols, C_4-dioxane, and esters of methacrylic, alkanoic, pentenoic, and acetic acids [8, 9].

LTA cannot be used for mineral or metallic abrasives, which are most commonly used in heavy industrial blasting of steelwork. However, the lighter abrasives required for cleaning aluminium are possible candidates for LTA. Further work would be required to identify the VOCs generated by a particular abrasive medium before the technique could be recommended.

5.2.3 Acid Extraction and Digestion

Acid extraction and digestion is a multistage process that involves extracting metal contaminants from spent blasting debris into an acidic solution, separating the (solid) spent debris from the solution, and then precipitating the metal contaminants as metal salts. After this process, the blasting debris is considered decontaminated and can be deposited in a landfill. The metals in the abrasive debris — now in the precipitate — are still hazardous waste but are of greatly reduced volume.

Trials of this technique were performed by the U.S. Army on spent, contaminated coal slag; mixed plastic; and glass bead abrasives. Various digestive processes and acids were used, and leachable metal concentrations of lead, cadmium, and chromium were measured using the TCLP method before and after the acid digestion. The results were disappointing: the acid digestion processes removed only a fraction of the total heavy metal contaminants in the abrasives [9]. Based on these results, this technique does not appear to be promising for treating spent abrasive.

5.3 METHODS FOR STABILIZING LEAD

Stabilizing lead means treating the paint debris so that the amount of lead leaching out is lowered, at least temporarily. There are concerns about both the permanence and effectiveness of these treatments. The major stabilization methods are explained in this section.

5.3.1 STABILIZATION WITH IRON

Iron (or steel) can stabilize lead in paint debris so that the rate at which it leaches out into water is greatly reduced. Generally, 5% to 10% (by weight) of iron or steel abrasive added to a nonferrous abrasive is believed to be sufficient to stabilize most pulverized lead paints [1].

The exact mechanism is unknown, but one reasonable theory holds that the lead dissolves into the leachate water but then immediately plates out onto the steel or iron. The lead ions are reduced to lead metal by reaction with the metallic iron [5], as shown here:

$$Pb^{2+} \quad + \quad Fe^0 \quad \rightarrow \quad Pb^0 \quad + \quad Fe^{2+}$$
$$\text{(ion)} \quad\quad \text{(metal)} \quad\quad\quad \text{(ion)} \quad\quad \text{(metal)}$$

The lead metal is not soluble in the acetic acid used for extracting metals in the TCLP test (see Section 5.1.2); therefore, the measured soluble lead is reduced. Bernecki et al. [10] make the important point that iron stabilizes only the lead at the exposed surface of the paint chips; the lead inside the paint chip, which comprises most of it, does not have a chance to react with the iron. Therefore, the polymer surrounding the lead pigment may break down over time in the landfill, allowing the bulk lead to leach out. The size of the pulverized paint particles is thus critical in determining how much of the lead is stabilized; small particles mean that a higher percentage of lead will be exposed to the iron.

The permanency of the stabilization is an area of concern when using this technique. Smith [11] has investigated how long the iron stabilizes the lead. The TCLP extraction test was performed repeatedly using paint chips, coal slag abrasive, and 6% steel grit. Initially, the amount of lead leached was 2 mg/L; by the eighth extraction, however, the lead leaching out had increased to above the permitted 5 mg/L. In another series of tests, a debris of spent abrasive and paint particles (with no iron or steel stabilization) had an initial leaching level of 70 mg/L. After steel grit was added, the leachable lead dropped to below 5 mg/L. The debris was stored for six months, with fresh leaching solution periodically added (to simulate landfill conditions). After six months, the amount of lead leached had returned to 70 mg/L. These tests suggest that stabilization of lead with steel or iron is not a long-term solution.

The U.S. EPA has decided that this is not a practical treatment for lead. In an article in the March 1995 issue of the *Federal Register* [12], "The Addition of Iron Dust to Stabilize Characteristic Hazardous Wastes: Potential Classification as Impermissible Dilution," the issue is addressed by the EPA as follows:

While it is arguable that iron could form temporary, weak, ionic complexes...so that when analyzed by the TCLP test the lead appears to have been stabilized, the Agency believes that this "stabilization" is temporary, based upon the nature of the complexing. In fact, a report prepared by the EPA on *Iron Chemistry in Lead-Contaminated Materials* (Feb. 22 1994), which specifically addressed this issue, found that iron-lead bonds are weak, adsorptive surface bonds, and therefore not likely to be permanent. Furthermore, as this iron-rich mixture is exposed to moisture and oxidative conditions over time, interstitial water would likely acidify, which could potentially reverse any temporary stabilization, as well as increase the leachability of the lead.... Therefore, the addition of iron dust or filings to...waste...does not appear to provide long-term treatment.

5.3.2 STABILIZATION OF LEAD THROUGH pH ADJUSTMENT

The solubility of many forms of lead depends on the pH of the water or leaching liquid. Hock and colleagues [13] have measured how much lead from white pigment can leach at various pH values using the TCLP test. The results are shown in Figure 5.1.

It is possible to add chemicals, for example calcium carbonate, to the blasting medium prior to blasting or to the debris afterward, so that the pH of the test solution in the TCLP is altered. At the right pH, circa 9 in the figure above, lead is not soluble in the test solution and thus is not measured. The debris "passes" the test for lead. However, this is not an acceptable technique because the lead itself is not permanently

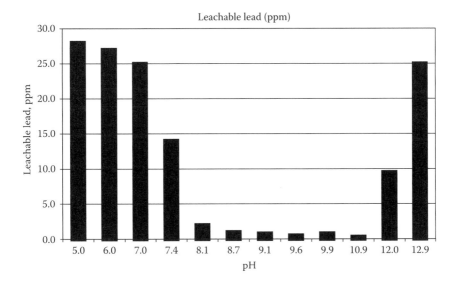

FIGURE 5.1 White lead leachability as a function of pH.
Source: Hock, V. et al., *Demonstration of lead-based paint removal and chemical stabilization using blastox,* Technical Report 96/20, U.S. Army Construction Engineering Research Laboratory, Champaign, IL, 1996.

stabilized. The effect, nonsoluble lead, is extremely temporary; after a short time, it leaches precisely as if no treatment had been done [13].

5.3.3 STABILIZATION OF LEAD WITH CALCIUM SILICATE AND OTHER ADDITIVES

5.3.3.1 Calcium Silicate

Bhatty [14] has stabilized solutions containing salts of cadmium, chromium, lead, mercury, and zinc with tricalcium silicate. Bhatty proposes that, in water, tricalcium silicate becomes calcium silicate hydrate, which can incorporate in its structure metallic ions of cadmium and other heavy metals.

Komarneni and colleagues [15–17] have suggested that calcium silicates exchange Ca^{2+} in the silicate structure for Pb^{2+}. Their studies have shown that at least 99% percent of the lead disappears from a solution as a lead-silicate-complex precipitate.

Hock and colleagues [13] have suggested a more complex mechanism to explain why cement stabilizes lead: the formation of lead carbonates. When cement is added to water, the carbonates are soluble. Meanwhile, the lead ions become soluble because lead hydroxides and lead oxides dissociate. These lead ions react with the carbonates in the solution and precipitate as lead carbonates, which have limited solubility. Over time, the environment in the concrete changes; the lead carbonates dissolve, and lead ions react with silicate to form an insoluble, complex lead silicate. The authors point out that no concrete evidence supports this mechanism; however, it agrees with lead stabilization data in the literature.

5.3.3.2 Sulfides

Another stabilization technique involves adding reactive sulfides to the debris. Sulfides — for example, sodium sulfide — react with the metals in the debris to form metal sulfides, which have a low solubility (much lower, for example, than metal hydroxides). Lead, for example, has a solubility of 20 mg/liter as a hydroxide, but only 6×10^{-9} mg/liter as a sulfide [18].

If the solubility of the metal is reduced, the leaching potential is then also reduced. Robinson [19] has studied sulfide precipitation and hydroxide precipitation of heavy metals, including lead, chromium, and cadmium; he saw less leaching among the sulfides, which also had lower solubility. Robinson also reported that certain sulfide processes could stabilize hexavalent chromium without reducing it to trivalent chromium (but does not call it sulfide precipitation and does not describe the mechanism). Others in the field have not reported this.

Means and colleagues [20] have also studied stabilization of lead and copper in blasting debris with sulfide agents and seen that they could effectively stabilize lead. They make an important point: that mechanical–chemical form of a pulverized paint affects the stabilization. The sulfide agent is required to penetrate the polymer around the metal before it can react with and chemically stabilize the metal. In their research, Means and colleagues used a long mixing time in order to obtain the maximum stabilization effect.

5.4 DEBRIS AS FILLER IN CONCRETE

Solidification of hazardous wastes in portland cement is an established practice [18]; it was first done in a nuclear waste field in the 1950s [4]. Portland cement has several advantages:

- It is widely available, inexpensive, and of fairly consistent composition everywhere.
- Its setting and hardening properties have been extensively studied.
- It is naturally alkaline, which is important because the toxic metals are less soluble at higher pH levels.
- Leaching of waste in cement has been extensively studied.

Portland cement has one major disadvantage: some of the chemicals found in paint debris have a negative effect on the set and strength development of the cement. Lead, for example, retards the hydration of portland cement. Aluminum reacts with the cement to produce hydrogen gas, which lowers the strength and increases permeability of the cement [4]. Some interesting work has been done, however, in adding chemicals to the cement to counteract the effects of lead and other toxic metals.

The composition of portland cement implies that, in addition to solidification, stabilization of at least some toxic metals is taking place.

5.4.1 PROBLEMS THAT CONTAMINATED DEBRIS POSE FOR CONCRETE

Hydration is the reaction of portland cement with water. The most important hydration reactions are those of the calcium silicates, which react with water to form calcium silicate hydrate and calcium hydroxide. Calcium silicate hydrate forms a layer on each cement grain. The amount of water present controls the porosity of the concrete: less water results in a denser, stronger matrix, which in turn leads to lower permeability and higher durability and strength [21].

Lead compounds slow the rate of hydration of portland cement; as little as 0.1% (w/w) lead oxide can delay the setting of cement [22]. Thomas and colleagues [23] have proposed that lead hydroxide precipitates very rapidly onto the cement grains, forming a gelatinous coating. This acts as a diffusion barrier to water, slowing — but not stopping — the rate at which it contacts the cement grains. This model is in agreement with Lieber's observations that the lead does not affect the final compressive strength of the concrete, merely the setting time [22]. Shively and colleagues [24] observed that the addition of wastes containing arsenic, cadmium, chromium, and lead had a delay before setting when mixed with portland cement, but the wastes' presence had no effect on final compressive strength of the mortar. Leaching of the toxic metals from the cement was greatly reduced compared with leaching from the original (untreated) waste. The same results using cadmium, chromium, and lead were seen by Bishop [25], who proposed that cadmium is adsorbed onto the pore walls of the cement matrix, whereas lead and chromium become insoluble silicates bound into the matrix itself. Many researchers have found that additives, such as sodium silicate, avoid the delayed-set problem; sodium silicate is believed to either form low-solubility metal oxide/silicates or possibly encapsulate the metal ions in silicate- or metal-silicate gel matrices. Either way, the metals are removed from solution before they precipitate on the cement grains.

Compared with lead, cadmium and chromium have negligible effects on the hardening properties of portland cement [26, 27].

5.4.2 ATTEMPTS TO STABILIZE BLASTING DEBRIS WITH CEMENT

The University of Texas at Austin has done a large amount of research on treatment of spent abrasive media by portland cement. Garner [28] and Braband [29] have studied the effects of concrete mix ingredients, including spent abrasives and counteracting additives, on the mechanical and leaching properties (TCLP) of the resulting concrete. They concluded that it is possible to obtain concrete using spent abrasive with adequate compressive strength, permeability resistance, and leaching resistance. Some of their findings are summarized here:

- The most important factors governing leaching, compressive strength, and permeability were the water/cement ratio and the cement content. In general, as the water/cement ratio decreased and the cement content increased, leaching decreased and compressive strength increased.
- As the contamination level of a mix increased, compressive strength decreased. (It should be noted that this is not in agreement with Shively's [24] results [see section above].)
- Mixes with lower permeability also had lower TCLP leaching concentrations.
- Mixing sequence and time were important for the success of the concrete. Best performance was obtained by thoroughly mixing the dry components prior to adding the liquid components. It was necessary to mix the mortar for a longer period than required for ordinary concrete to ensure adequate homogenization of the waste throughout the mix.
- Set times and strength development became highly unpredictable as the contamination level of the spent abrasives increased.
- Contamination level of the spent abrasives was variable. Possible factors include the condition and type of paint to be removed, the type of abrasive, and the type of blasting process. These factors contribute to the particle size of the pulverized paint and its concentration in the spent blasting abrasives.
- No relationship was found between the leaching of the individual metals and the concrete mix ingredients.

Salt and colleagues [4] have investigated using accelerating additives to counteract the effects of lead and other heavy metals in the spent abrasive on the set, strength, and leaching of mortars made with portland cement and used abrasive debris. Some of their findings are summarized here:

- Sodium silicate was most effective in reducing the set time of portland cement mixed with highly contaminated debris, followed by silica fume and calcium chloride. Calcium nitrite was ineffective at reducing the set time for highly contaminated wastes.

- The combination of sodium silicate and silica fume provided higher compressive strength and lower permeability than separate use of these compounds.
- The set time is proportional to the lead/portland cement ratio; decreasing the ratio decreases the set time. The compressive strength of the mortar, on the other hand, is inversely proportional to this ratio.
- For the most highly contaminated mixtures, accelerators were required to achieve setting.
- All the mortars studied had TCLP leaching concentrations below EPA limits. No correlation between the types and amounts of metals in the wastes and the TCLP leaching results was found.
- The proper accelerator and amounts of accelerator necessary should be determined for each batch of blasting debris (where the batch would be all the debris from a repainting project, which could be thousands of tons in the case of a large structure) by experimenting with small samples of debris, accelerators, and portland cement.

Webster and colleagues [30] have investigated the long-term stabilization of toxic metals in portland cement by using sequential acid extraction. They mixed portland cement with blasting debris contaminated with lead, cadmium, and chromium. The solid mortar was then ground up and subjected to the TCLP leaching test. The solid left after filtering was then mixed with fresh acetic acid and the TCLP test was repeated. This process was done sequentially until the pH of the liquid after leaching and filtering was below 4. Their findings are summarized here:

- The amount of lead leaching was strongly dependent on the pH of the liquid after being mixed with the solid; lead with a pH below 8 began to leach, and the amount of lead leaching rose dramatically with each sequential drop in pH.
- Cadmium also began leaching when the pH of the liquid after being mixed with the solid dropped below 8; the amount leaching reached a maximum of 6 and then fell off as the pH continued to drop. This could be an artificial maximum, however, because the amount of cadmium was low to begin with; it could be that by the time the pH had dropped to 5, almost all the cadmium in the sample had leached out.
- The authors suggest that the ability of the calcium matrix to resist breakdown (due to acidification) in the concrete is important for the stabilization of lead and cadmium.
- Chromium began leaching with the first extractions (pH 12); the amount leached was constant with each of the sequential extractions until the pH dropped below 6. Because the amount of chromium in the debris was also low, the authors suggest that chromium has no pH dependency for leaching; instead it merely leaches until it is gone. This finding was supported

by the fact that, as the chromium concentration in the blasting debris increased, the TCLP chromium concentration also increased.

• The authors noted that the sequential acid leaching found in their testing was much more harsh than concrete is likely to experience in the field; however, it does hint that stabilization of toxic metals with portland cement will work only as long as the concrete has not broken down.

5.4.3 PROBLEMS WITH ALUMINUM IN CONCRETE

Not all metals can be treated with portland cement alone; aluminum in particular can be a problem. Khosla and Leming [31] investigated treatment of spent abrasive containing both lead and aluminum by portland cement. They found that aluminum particles corroded rapidly in the moist, alkaline environment of the concrete, forming hydrogen gas. The gas caused the concrete to expand and become porous, decreasing both its strength and durability. No feasible rapid-set (to avoid expansion) or slow-set (to allow for corrosion of the aluminum while the concrete was still plastic) was found in this study. (Interestingly, the amount of lead leaching was below the EPA limit despite the poor strength of the concrete.) However, Berke and colleagues [32] found that calcium nitrate was effective at delaying and reducing the corrosion of aluminum in concrete.

5.4.4 TRIALS WITH PORTLAND CEMENT STABILIZATION

In Finland, an on-site trial has been conducted of stabilization of blasting debris with portland cement. The Koria railroad bridge, approximately 100 m long and 125 years old, was blasted with quartz sand. The initial amount of debris was 150 tons. This debris was run through a negative-pressure cyclone and then sieved to separate the debris into four classes. The amount of "problem debris" — defined in this pilot project as debris containing more than 60 mg of water-soluble heavy metals per kilogram debris —remaining after the separation processes was only 2.5 tons. This was incorporated into the concrete for bottom plates at the local disposal facility [33].

The U.S. Navy has investigated ways to reduce slag abrasive disposal costs in shipyards and found two methods that are both economically and technically feasible: reusing the abrasive and stabilizing spent abrasive in concrete. In this investigation, copper slag abrasive picked up a significant amount of organic contamination (paint residue), making it unsuitable for portland cement concrete, for which strength is a requirement. It was noted, however, that the contaminated abrasive would be suitable in asphalt concrete [34].

5.5 OTHER FILLER USES

Blasting debris can also be incorporated as filler into asphalt and bricks. Very little is reported in the literature about these uses, in particular which chemical forms the heavy metals take, how much leaching occurs, and how permanent the whole arrangement is. In Norway, one company, Per Vestergaard Handelsselkab, has reported sales of spent blasting media for filler in asphalt since 1992 and for filler in brick since 1993. They report that variations in the quality (i.e., contamination levels) of the debris have been a problem [35].

REFERENCES

1. Trimber, K., *Industrial Lead Paint Removal Handbook*, SSPC Publication 93-02, Steel Structures Painting Council, Pittsburgh, PA, 1993, 152.
2. Appleman, B.R., *Bridge paint: Removal, containment, and disposal*, Report 175, National Cooperative Highway Research Program, Transportation Research Board, Washington DC, 1992.
3. Harris, J. and Fleming, J., Testing lead paint blast residue to pre-determine waste classification, in *Lead Paint Removal From Industrial Structures*, Steel Structures Painting Council, Pittsburgh, PA, 1989, 62.
4. Salt, B. et al., *Recycling contaminated spent blasting abrasives in portland cement mortars using solidification/stabilization technology*, Report CTR 0-1315-3F, U.S. Department of Commerce, National Technical Information Service (NTIS), Springfield, 1995.
5. *Test methods for evaluating solid waste, physical/chemical methods*, 3rd ed., EPA Publication SW-846, GPO doc. no. 955-001-00000-1, Government Printing Office, Washington DC, 1993, Appendix II-Method 1311.
6. Drozdz, S., Race, T. and Tinklenburg, K., *J. Prot. Coat. Linings*, 17, 41, 2000.
7. Tapscott, R.E., Blahut, G.A. and Kellogg, S.H., *Plastic media blasting waste treatments*, Report ESL-TR-88-122, Engineering and Service Laboratory, Air Force Engineering and Service Center, Tyndall Air Force Base, FL, 1988.
8 Jermyn, H. and Wichner, R.P., Plastic media blasting (PMB) waste treatment technology, in *Proc. Air and Waste Management Conference*, Air and Waste Management Association, Vancouver, 1991, Paper 91-10-18
9. Boy, J. H., Rice, T.D. and Reinbold, K.A. *Investigation of separation, treatment, and recycling options for hazardous paint blast media waste*, USACERL Technical Report 96/51, Construction Engineering Research Laboratories, U.S. Army Corps of Engineers, Champaign, IL, 1996.
10. Bernecki, T.F., et al., *Issues impacting bridge painting: An overview*, FHWA/RD/94/098, U.S. Federal Highway Administration, National Technical Information Service, Springfield, 1995.
11. Smith, L., oral presentation, *Sixth Annual Conference on Lead Paint Removal and Abatement*, Steel Structures Painting Council, Pittsburgh, PA, 1993.
12. U.S.A. Federal Register, Vol. 60, No. 41, National Archive and Record Administration, Washington D.C., 1995.
13. Hock, V. et al., *Demonstration of lead-based paint removal and chemical stabilization using blastox*, Technical Report 96/20, U.S. Army Construction Engineering Research Laboratory, Champaign, IL, 1996.
14. Bhatty, M., Fixation of metallic ions in portland cement, in *Proc. National Conference on Hazardous Wastes and Hazardous Materials*, Washington DC, 1987, p. 140-145.
15. Komarneni, S., Roy, D. and Roy, R., *Cement Concrete Res.*, 12, 773, 1982.
16. Komarneni, S., *Nucl. Chem. Waste Manage.*, 5, 247, 1985.
17. Komarneni, S. et al., *Cement Concrete Res.*, 18, 204, 1988.
18. Conner, J.R., *Chemical Fixation and Solidification of Hazardous Wastes,* van Nostrand Reinhold, New York, 1990.

19. Robinson, A.K., Sulfide vs. hydroxide precipitation of heavy metals from industrial wastewater, in *Proc. First Annual Conference on Advanced Pollution Control for the Metal Finishing Industry*, Report EPA-600/8-78-010, Environmental Protection Agency, Washington DC, 1978, 59.

20. Means, J. et al., in *Engineering Aspects of Metal-Waste Management*, Iskandar, I.K. and Selim, H.M., Eds., Lewis Publishers, Chelsea, MI, 1992, p. 199.

21. Mindess, S. and Young, J., *Concrete*, Prentice-Hall, Inc., Englewood Cliffs, 1981.

22. Lieber, W., Influence of lead and zinc compounds on the hydration of portland cement, in *Proc, 5th International Symposium on the Chemistry of Cements*, Tokyo, 1968, Vol. 2, p. 444.

23. Thomas, N., Jameson, D., and Double, D., *Cement Concrete Res.*, 11, 143, 1981.

24. Shively, W. et al., *J. Water Pollut. Control Fed.*, 58, 234, 1986.

25. Bishop, P., *Hazardous Waste Hazardous Mat.*, 5, 129, 1988.

26. Tashiro, C. et al., *Cement Concrete Res.*, 7, 283, 1977.

27. Cartledge, F. et al., *Environ. Science Technol.*, 24, 867, 1990.

28. Garner, A., *Solidification/stabilization of contaminated spent blasting media in portland cement concretes and mortars*, Master of Science thesis, University of Texas at Austin, Austin, TX, 1992.

29. Brabrand, D., *Solidification/stabilization of spent blasting abrasives with portland cement for non-structural concrete purposes*, Master of Science thesis, University of Texas at Austin, Austin, TX, 1992.

30. Webster, M. et al., *Solidification/stabilization of used abrasive media for non-structural concrete using portland cement*, NTIS Nr. Pb96-111125 (FHWA/TX-95-1315-2), U.S. Department of Commerce, National Technical Information Service, Springfield, 1994.

31. Khosla, N. and Leming, M., Recycling of lead-contaminated blasting sand in construction materials, in *Proc. SSPC Lead Paint Removal Symposium*, Steel Structures Painting Council, Pittsburgh, PA, 1988.

32. Berke, N., Shen, D. and Sundberg, K., Comparison of the polarization resistance technique to the macrocell corrosion technique, in *Corrosion Rates of Steel in Concrete*, ASTM STP 1065, American Society for Testing and Materials, Philadelphia, PA, 1990, 38.

33. Sörensen, P., *Vanytt*, 1, 26, 1996. (In Swedish)

34. *Final report, Feasibility and economics study of the treatment, recycling and disposal of spent abrasives*, Project N1-93-1, NSRP # 0529, National Shipbuilding Research Program, ATI Corp., Charleston, SC, 1999.

35. Bjorgum, A., *Behandling av avfall fra bläserensing. Del 3. Oppsummering av utredninger vedrorende behandling av avfall fra blåserensing*, Report Nr. STF24 A95326, SINTEF, Trondheim, 1995. (In Norwegian)

6 Weathering and Aging of Paint

This chapter presents a brief overview of the major mechanisms that cause aging, and subsequent failure, of organic coatings. Even the best organic coatings, properly applied to compatible substrates, eventually age when exposed to weather, losing their ability to protect the metal.

In real-life environments, the aging process that leads to coating failure can generally be described as follows:

1. Weakening of the coating by significant amounts of bond breakage within the polymer matrix. Such bond breakage may be caused chemically (e.g., through hydrolysis reactions, oxidation, or free-radical reactions) or mechanically (e.g., through freeze–thaw cycling, which leads to alternating tensile and compressive stresses in the coating).
2. Overall barrier properties may be decreased as bonds are broken in the polymeric backbone — in other words, as transportation of water, oxygen, and ions through the coating increases. The polymeric network may be plasticized by absorbed water, which softens it and makes it more vulnerable to mechanical damages. The coating may begin to lose small, water-soluble components, causing further damage. Flaws such as microcracks develop or, if preexisting, are enlarged in the coating.
3. Even more transportation of water, oxygen, and ions through the coating.
4. Deterioration of coating-metal adhesion at this interface.
5. Development of an aqueous phase at the coating/metal interface.
6. Activation of the metal surface for the anodic and cathodic reactions.
7. Corrosion and delamination of the coating.

Many factors can contribute in various degrees to coating degradation, such as:

- Ultraviolet (UV) radiation
- Water and moisture uptake
- Elevated temperatures
- Chemical damage (e.g., from pollutants)
- Thermal changes
- Molecular and singlet oxygen
- Ozone
- Abrasion or other mechanical stresses

The major weathering stresses that cause degradation of organic coatings are the first four in the list above: UV radiation, moisture, heat, and chemical damage. And, of course, interactions between these stresses are to be expected; for example, as the polymeric backbone of a coating is slowly being broken down by UV light, the coating's barrier properties can be expected to worsen. Ranby and Rabek [1] have shown that under UV stress, polyurethanes react with oxygen to form hydroperoxides and that this reaction is accelerated by water. Another example is the temperature-condensation interaction. Elevated temperatures by themselves can damage a polymer; however, they can also create condensation problems, for example, if high daytime temperatures are followed by cool nights. These day/night (diurnal) variations in temperature determine how much condensation occurs, as the morning air warms up faster than the steel.

Various polymers, and, therefore, coating types, react differently to changes in one or more of these weathering stresses. In order to predict the service life of a coating in a particular application, therefore, one must know not only the environment — average time of wetness, amounts of airborne contaminants, UV exposure, and so on — but also how these weathering stresses affect the particular polymer [2].

6.1 UV BREAKDOWN

Sunlight is the worst enemy of paint. It is usually associated with aesthetic changes, such as yellowing, color change or loss, chalking, gloss reduction, and lowered distinctness of image. More important than the aesthetic changes, however, is the chemical breakdown and worsened mechanical properties caused by sunlight. The range of potential damage is enormous [3-7] and includes:

- Embrittlement
- Increased hardness
- Increased internal stress
- Generation of polar groups at the surface, leading to increased surface wettability and hydrophilicity
- Changed solubility and crosslink density

In terms of coating performance, this translates into alligatoring, checking, crazing, and cracking; decreased permeation barrier properties; loss of film thickness; and delamination from the substrate or underlying coating layer.

All the damage described above is created by the UV component of sunlight. UV light is a form of energy. When this extra energy is absorbed by a chemical compound, it makes bonds and break bonds. Visible light does not contain the energy required to break the carbon–carbon and carbon–hydrogen bonds most commonly found on the surface of a cured coating. However, just outside of the visible range light in the wavelength range of 285 to 390 nm contains considerably more energy, commonly enough to break bonds and damage a coating. The 285 to 390 nm range causes almost all weathering-induced paint failure down at ground level [4]. At the short end of the UV range, we find the most destructive radiation. The damage caused by short-wave radiation is limited, though, to the topmost surface layers of the coating. Longer wave UV radiation penetrates the film more deeply, but causes less damage [8-10]. This leads to an inhomogeneity in the coating, where the top surface can be more highly crosslinked than the bulk of the coating layer [4]. As the top surface of the film eventually breaks up, chalking and other degradation phenomena become apparent. (The light located below 285 nm, with even higher energy, can easily break carbon-carbon and carbon-hydrogen bonds and has enough energy left over for considerably more mischief as well. However, Earth's atmosphere absorbs most of this particular wavelength band of radiation and, therefore, it is a concern only for aircraft coatings, which receive less protection from the ozone layer.)

The interactions of coatings with UV radiation may be broadly classed as follows:

- Light is reflected from the film.
- Light is transmitted through the film.
- Light is absorbed by a pigment or by the polymer.

In general, reflectance and transmittance pose no threats to the lifespan of the coating. Absorption is the problem. When energy from the sun is absorbed, it leads to chemical destruction (see Section 6.1.3).

6.1.1 REFLECTANCE

Light is reflected from the film by the use of leafy or plate-like metal pigments located at the top of the coating. These are surface-treated so that the binder solution has difficulty wetting them. When the film is applied, the plate-like pigments float to the top of the wet film and remain there throughout the curing process. The dried film has a very thin layer of binder on top of a layer of pigment that is impermeable to light. The binder on top of the pigment layer may be broken down by UV radiation and disappear; but as long as the leafy pigments can be held in place, the bulk of the binder behind the leafy pigments are shielded from sunlight.

6.1.2 TRANSMITTANCE

Transmitted light, which passes through the film without being absorbed, does not affect the structure of the film. Of course, if a coating layer underneath is sensitive to UV radiation, problems can occur. Epoxy coatings, which are the most important class of anticorrosion primer, are highly sensitive to UV radiation. These primers are generally covered by a topcoat whose main function is to not transmit the UV radiation.

6.1.3 ABSORPTION

Light can be absorbed by a pigment, the binder, or an additive. Light absorbed by the pigment is dissipated as heat, which is a less destructive form of energy than UV light [4]. The real damage comes from the UV radiation absorbed by the nonpigment components of the coating — that is, the polymeric binder or additive.

UV energy absorbed by the binder or additive can wreak havoc in wild ways. The extra energy can go into additional crosslinking of the polymer, or it can start breaking the existing bonds.

Because the polymer chains in the cured film are well anchored and already crosslinked, further crosslinking results in additional tightening of the polymer chains [7]. This increases the internal stress of the cured film, which in turn leads to hardening, decreased flexibility, and embrittlement. If the internal stresses overcome the cohesive strength of the film, then the unfortunate end is cracking; if failure takes the form of lost adhesion at the coating/metal interface, then delamination is seen. Both, of course, can happen simultaneously.

Instead of causing additional crosslinking, the UV energy could break bonds in the polymer or another component of the coating. Free radicals are thus initiated. These free radicals react with either:

- Oxygen to produce peroxides, which are unstable and can react with polymer chains
- Other polymer chains or coating components to propagate more free radicals

Reaction of the polymer chain with peroxides or free radicals leads to chain breaking and fragmentation. "Scissoring," a term used to describe this reaction, is an apt description. The effect is exactly as if a pair of scissors was let loose inside the coating, cutting up the polymer backbone. The destruction is enormous. When scissoring cuts off small molecules, they can be volatilized and make their way out of the coating. The void volume necessarily increases as small parts of the binder disappear (and, of course, ultimately the film thickness decreases). The internal stress on the remaining anchored polymer chains increases, leading to worsened mechanical properties. After enough scissoring, the crosslink density has been significantly altered for the worse, loss of film thickness occurs, and a decrease in permeation barrier properties is seen. The destruction stops only when two free radicals combine with each other, a process known as *termination* [4, 11].

Table 6.1 summarizes the effects on the coating when absorbed UV energy goes into additional crosslinking, scissoring or generating polar groups at the coating surface.

TABLE 6.1
Effects of Absorbed UV Energy

Absorbed UV energy goes into...	...which causes	...and ultimately
Additional crosslinking	Increased internal stress, leading to hardening, decreased flexibility, and eventually embrittlement	Cracking, delamination, or both
Scissoring	Increased internal stress Increased void volume Worsened crosslink density	Loss of film thickness Decrease of permeation barrier properties
Generation of polar groups at the surface	Increased surface wettability and hydrophilicity	Decrease of permeation barrier properties

Ideally, selection of binders that absorb little or no UV radiation should minimize the potential damage from this source. In reality, however, even paints based on these binders can prove vulnerable because other components — both those intentionally added and those that were not — often compromise the coating as a whole. Components that can be said to have been added intentionally are, of course, pigments and various types of additives: antiskinning, antibacterial, emulsifying, colloid-stabilizing, flash-rust preventing, flow-controlling, thickening, viscosity-controlling, additives, *ad infinitum*. Examples of unintentional components are catalysts or monomer residues left over from the polymer processing; these may include groups that are highly reactive in the presence of UV radiation, such as ketones and peroxides. Interestingly, impurities can sometimes show a beneficial effect. When studying waterborne acrylics, Allen and colleagues [12] have found that low levels of certain comonomers reduced the rate of hydroperoxidation. The researchers speculate that the styrene comonomer reduced the unzipping reaction that the UV otherwise would cause.

6.2 MOISTURE

Moisture (water or water vapor) can come from several sources, including water vapor in the surrounding air, rain, and condensation as temperatures drop at night. Paint films constantly absorb and desorb water to maintain equilibrium with the amount of moisture in the environment. Water is practically always present in the coating. In a study of epoxy, chlorinated rubber, alkyd and linseed oil paints, Lindqvist [13] found that even in stagnant air at 25°C and 20% relative humidity (RH), the smallest equilibrium amount of water measured was 0.04 wt %.

Water or water vapor is taken up by the coating as a whole through pores and microcracks; the binder itself also absorbs moisture. Water uptake is not at all homogeneous; it enters the film in several different ways and can accumulate in various places [13, 14]. Within the polymer phase, water molecules can be randomly distributed or aggregate into clusters, can create a watery interstice between binder and pigment particle, can exist in pores and voids in the paint film, and can accumulate at the metal-coating interface. Once corrosion has begun, water can exist in blisters or in corrosion products at the coating-metal interface.

Water molecules can exist within the polymer phase because polymers generally contain polar groups that chemisorb water molecules. The chemisorbed molecules can be viewed as bound to the polymer because the energy for chemisorption (10 to 100 kcal/mole) is similar to that required for chemical bonding. The locked, chemisorbed molecule can be the center for a water cluster to form within the polymer phase [13].

When water clusters form in voids or defects in the film, they can behave as fillers, stiffening the film and causing a higher modulus than when the film is dry. Funke and colleagues [14] concluded that moisture in the film can have seemingly contradictory effects on the coating's mechanical properties because several different — and sometimes opposite — phenomena are simultaneously occurring.

Two of the most important parameters of water permeation are solubility and diffusion. Solubility is the maximum amount of water that can be present in the coating in the dissolved state. Diffusion is how mobile the water molecules are in the coating [15]. The permeability coefficient, P, is the product of the diffusion coefficient, D, and the solubility, S [16]:

$$P = D \times S$$

In accelerated testing, the difference in absorption and desorption rates of water for various coatings is also important (see Chapter 7).

The uptake of water affects the coating in several ways [17]:

- Chemical breakdown
- Weathering interactions
- Hygroscopic stress
- Blistering/adhesion loss

6.2.1 CHEMICAL BREAKDOWN

Water is an excellent solvent for atmospheric contaminants, such as salts, sulfites, and sulphates. Airborne contaminants would probably never harm coated metals, if not for the fact that they so easily become Cl^- or SO_4^{2-} ions in water. The water and ions, of course, fuel corrosion beneath the coating.

Water can also be a solvent for some of the additives in the paint, causing them to dissolve or leach out of the cured film. And finally, it can act as a plasticizer in the polymeric network, softening it and making it more vulnerable to mechanical damages. Lefebvre and colleagues [18], working with epoxy films, have proposed that each coating had a critical RH. Above the critical RH, water condensed on the OH groups of the polymer, breaking interchain hydrogen bonds and displacing adsorbed OH groups from the substrate surface. The loss of adhesion resulting from this was reversible. However, an irreversible effect was the reaction of the water with residual oxirane rings in the coating to form diols. This led to an irreversible increase in solubility and swelling of the film.

6.2.2 WEATHERING INTERACTIONS

As previously noted, the major weathering stresses interact with each other. Perera and colleagues have shown that temperature effects are inseparable from the effects of water [19, 20]. The same is even more true for chemical effects (see Section 6.4).

The effects of UV degradation can be worsened by the presence of moisture in the film [1]. As a binder breaks down due to UV radiation, water-soluble binder fragments can be created. These dissolve when the film takes up water, are removed from the film upon drying, and add to the decrease in film density or thickness.

6.2.3 HYGROSCOPIC STRESS

This section focuses on the changes in the coating's internal stresses — both tensile and compressive — caused by wetting and drying the coating. As a coating takes up water, it swells, causing compressive stresses in the film. As the coating dries, it contracts, causing tensile stress. These compression and tension forces have adverse effects on the film's cohesive integrity and on its adhesion to the substrate. Of the two types of stresses, the tensile stresses formed as the coating dries have the greater effect [9, 11, 21].

Coating stress is a dynamic phenomenon; it changes drastically during water uptake and desorption. Sato and Inoue [22] have reported that the initial tensile stresses (left over from shrinkage during film formation) of the dry film decrease to zero as moisture is absorbed. Once the initial tensile stresses have been negated by water uptake, further uptake leads to build-up of compressive stresses. If the film is dried, tensile (shrinkage) stresses redevelop, but to a lower degree than originally seen. Some degree of permanent creep was seen in Sato and Inoue's study; it was attributed to breaking and reforming valency associations in the epoxy polymer. The same trend of initial tensile stress reduction, followed by compressive stress build-up was seen by Perera and Vanden Eynde [23] with a polyurethane and a thermoplastic latex coating.

Hygroscopic stresses are interrelated with ambient temperature [11, 20]. They also depend heavily on the glass transition temperature (T_g) of the coating [24]. In immersion studies, Perera and Vanden Eynde examined the stress of an epoxy coating whose T_g was near — even below — the ambient temperature [25]. The films in question initially had tensile stress from the film formation. Upon immersion, this stress gradually disappeared. As in the previously cited studies, compressive stresses built up. The difference was that these stresses then dissipated over several days even though immersion continued. Hare also noted dissipation of compressive stresses as the difference between $T_{ambient}$ and T_g is reduced; he attributes it to a reduced modulus and a flexibilizing of the film [11]. Because of the low T_g of the film, stress relaxation occurred and the compressive stresses due to water uptake disappeared.

Hygroscopic stresses have a very real effect on coating performance. If a coating forms high levels of internal stress during cure — not uncommon in thick, highly crosslinked coatings — then applying other stresses during water uptake or desorption can lead to cracking or delamination. Hare has reported another problem: cases where the film expansion during water uptake created a strain beyond the film's yield point. Deformation here is irreversible; during drying, permanent wrinkles are left in the dried paint [17]. Perera has pointed out that hygroscopic stress can be critical to designing accelerated tests for coatings. For example, a highly crosslinked coating can undergo more damage in the few hours it dries after the salt-spray test has ended than it did in the entire time (hundreds of hours) of the test itself [26].

6.2.4 BLISTERING/ADHESION LOSS

Blistering is not, strictly speaking, brought about by aging of the coating. It would be more correct to say that blistering is a sign of failure of the coating-substrate system. Blistering occurs when moisture penetrates through the film and accumulates at the coating-metal interface in sufficient numbers to force the film up from the metal substrate. The two types of blistering in anticorrosion paints — alkaline and neutral — are caused by different mechanisms.

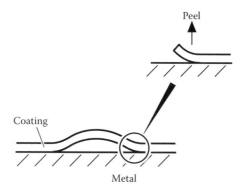

FIGURE 6.1 Peel forces at the edge of a blister

6.2.4.1 Alkaline Blistering

Alkaline blistering occurs when cations, such as sodium (Na^+), migrate along the coating-metal interface to cathodic areas via coating defects, such as pores or scratches. At the cathodic areas, the cations combine with the hydroxyl anions produced by corrosion to form sodium hydroxide (NaOH). The result is a strongly alkaline aqueous solution at the cathodic area. As osmotic forces drive water through the coating to the alkaline solution, the coating is deformed upward — a blister begins. At the coating-metal solution interface, the coating experiences peel forces, as shown in Figure 6.1. It is well established that the force needed to separate two adhering bodies is much lower in peel geometry than in the tensile geometry normally used in adhesion testing of coatings. At the edge of the blister, the coating may be adhering as tightly as ever to the steel. However, because the coating is forced upward at the blister, the coating at the edge is now undergoing peeling and the force needed to detach the coating in this geometry is lower than the forces measured in adhesion tests. This facilitates growth of the blisters until (probably) the solution is diluted with water and the osmotic forces have decreased.

Leidheiser and colleagues [27] have shown that cations diffuse laterally via the coating-metal interface, rather than through the coating. Their elegantly simple experiment demonstrating this is shown in Figure 6.2. Adhesion is significantly less under wet conditions (see "Wet Adhesion" in Chapter 1), making ion migration along the interface easier.

6.2.4.2 Neutral Blistering

Neutral blisters contain solution that is weakly acid to neutral. No alkali cations are involved. The first step is undoubtedly reduction of adhesion due to water clustering at the coating-metal interface. Funke [28] postulates that differential aeration is responsible for neutral blistering. The steel under the water does not have as ready access to oxygen as the adjacent steel, and polarization arises. The oxygen-poor center of the blister becomes anodic and the periphery is cathodic. Funke's mechanism of neutral blistering is shown in Figure 6.3.

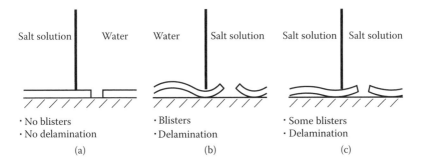

| Salt solution | Water | Water | Salt solution | Salt solution | Salt solution |

· No blisters   · Blisters   · Some blisters
· No delamination   · Delamination   · Delamination
      (a)   (b)   (c)

FIGURE 6.2 Experiment of Leidheiser et al. establishing route of cation diffusion
Source: Leidheiser, H., Wang, W., and Igetoft, L., *Prog. Org. Coat.,* 11, 19, 1983.

6.3 TEMPERATURE

In general, ambient temperature changes can alter:

- Balance of stresses in the coating/substrate system
- Mechanical properties of the viscoelastic coating
- Diffusion (usually of water) through the coating

The balance of stresses is affected in various ways. At slightly elevated temperatures, crosslinking of the polymer can continue far beyond what is desirable; the paint becomes too stiff and cracks with minimal amount of mechanical stress. Even if undesired crosslinking does not occur, bonds that are needed begin to break at higher temperatures, and the polymer is weakened. Differences in coefficients of thermal expansion also cause thermal stress; epoxies or alkyds, for example, typically

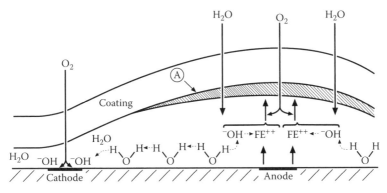

Ⓐ Secondary oxidation products

FIGURE 6.3 Mechanism of neutral blistering
Source: Funke, W., *Ind. Eng. Chem. Prod. Res. Dev.,* 24, 343, 1985

have a coefficient of thermal expansion that is twice that of aluminium or zinc and four times that of steel [29].

Another factor that must be considered at elevated temperatures is the glass transition temperature (T_g) of the polymer used in the binder. This is the temperature above which the polymer exists in a rubbery state and below which it is in the glassy state. Using coatings near the T_g range is problematic, because the binder's most important properties change in the transition from glassy to rubbery. For example, above the T_g, polymer chain segments undergo Brownian motion. Segments with appropriate functional groups for bonding are increasingly brought into contact with the metal surface. An increase in the number of bond sites can dramatically improve adhesion; wet adhesion in particular can be much better above the T_g than below it.

Increased Brownian motion is also associated with negative effects, such as increased diffusion. Above the T_g, the Brownian motion gives rise to the continuous appearance and disappearance of small pores, 1 to 5 nm or smaller, within the binder matrix. The size of these small pores compares to the "jump distance" of diffusing molecules — the distance that has to be covered by a molecule moving from one potential-energy minimum to a neighboring one in the activated diffusion process. The permeation rate through these small pores is linked to temperature to the same degree that the chain mobility is. That is, the chain mobility of elastomeric polymers shows a high degree of temperature dependence and thus favors activated diffusion at higher temperatures. As the crosslink density of the binder increases, segmental mobility decreases, even at elevated temperatures. Diffusion still occurs through large pore systems whose geometry is largely independent of temperature. The temperature dependence of diffusion in highly crosslinked binders is a result of the temperature dependence of the viscous flow of the permeating species.

Miszczyk and Darowicki have found that the increased water uptake at elevated temperatures can be to some extent irreversible; the absorbed water was not fully desorbed during subsequent temperature decreases. They speculate that the excess water may be permanently located in microcracks, microvoids, and local delamination sites [29].

6.4 CHEMICAL DEGRADATION

All breakdowns in polymers could, of course, be regarded as chemical degradation of some sort. What is meant here by the term "chemical degradation" is breakdown in the paint film that is induced by exposure to chemical contaminants in the atmosphere.

Atmospheric contaminants play a more minor role in polymer breakdown than do UV exposure, moisture, and (to a lesser degree) temperature. However, they can contribute to coating degradation, especially when they make the coating more vulnerable to degradation by UV light, water, or heat.

Mayne and coworkers have shown that the organic coating and the ions (e.g., sodium, potassium, calcium) in a solution interact, causing a gradual reduction in resistance of the coating (the "slow change," see Chapter 1, Section 1.2, "Protection Mechanisms of Organic Coatings") [30-34]. One interesting aspect is that, as long as it doesn't go too far, the reduction in resistance is reversible. The process is of course accelerated by heat; raising the temperature increases the amounts of ions exchanged, reducing the resistance and thus aging the coating. Coatings vary in their ability to resist this, but all exhibit the trend to some extent.

Chemical species, such as road salts and atmospheric contaminants in the wind or rain, are routinely deposited on paint surfaces. There, they combine with condensation to form aggressive, usually saline or acidic solutions. Most polymers used in modern coatings have good resistance to acids and salts; however, modern coatings also contain a large number of additives (see Chapter 2), which can prove vulnerable to chemical attack. For example, many coatings contain light stabilizers based on hindered amines to aid UV resistance. It is well known that the performance of these stabilizers is diminished by acids and pesticides [35]. When this occurs, chemical exposure makes the coating vulnerable to UV breakdown.

The number of studies in this area is limited, but a few have shown exposure of coatings to atmospheric contaminants to be detrimental. Sampers [35] reports that, in a study of polyolefin samples exposed both in Florida and on the Mediterranean coast of France, a dramatic difference was seen in polymer lifetime. Samples exposed on the Mediterranean had only half the life of those in Florida. The two stations had broadly similar weathering parameters; the differences should have led to *longer* lifetimes in France. Sampers concluded that constituents in the rain or wind had chemically interacted with the hindered amine light stabilizers in the polymers exposed in France, causing these samples to be especially vulnerable to UV degradation.

In a study of gloss retention of coatings exposed for 2 1/2 years at weathering sites in Kuwait, Carew and colleagues [36] reported a probable link between industrial pollution and coating damage, although in this case the damage seems to have been caused by dust from a cement factory. The sites in this study are described in Table 6.2. Because all the sites are located in the Shuaiba area of Kuwait's industrial belt, they should be very similar in temperature and humidity. The difference between sites is the distance from the Arabian Gulf and the amount and type of atmospheric pollution. Carew and colleagues found that coatings consistently showed the worst performance at site C, although this site is farther from the sea than sites A and B

TABLE 6.2
Description of Sites in Kuwait Study

Site	Distance from Sea	Pollution	Notes
A	0.2 km	Heavy	Downwind from refinery and salt and chlorine plant
B	0.55 km	Heavy	Next to refinery and desalination and electricity production plant
C	1.5 km	Heavy	Upwind from refinery, next to cement clinker factory
D	3 km	Mild	Rural area

Data from: Carew, J.A. et al., Weathering performance of industrial atmospheric coatings systems in the Arabian Gulf, *Proc. Corros. '94,* NACE, Houston, 1994, Paper 445.

and, being upwind, does not suffer from the refinery. However, they also noted heavy amounts of dust on the samples at this site, almost certainly from the cement factory next door. Analysis of the dust showed it to be similar to the composition of clinker cement. Cement, of course, is extremely alkaline, to which few polymers are resistant. At the high temperatures at these sites — up to 49°C — and with the very high amounts of water vapor available, soluble alkaline species in the dust deposits can form a destructive, highly alkaline solution that can break down cured binder. The extent to which the various coatings managed to retain gloss at this site is almost certainly a reflection of the polymer's ability to resist saponification.

In a study of coated panels exposed throughout two pulp and paper mills in Sweden, Rendahl and colleagues [37] found that the amounts of airborne H₂S and SO₂ at the various locations did not have a significant impact on coating performance. The effect of airborne chlorine in this study is not clear; the authors note that only total chlorine was measured, and the amounts of active corrosion-initiating species at each location are unknown.

Özcan and colleagues [38] examined the effects of very high SO_2 concentrations on polyester coatings. Using 0.286 atmosphere SO_2 (to simulate conditions in flue gases) and humidity ranging from 60% to 100% RH, they found that corrosion occurred only in the presence of water. At 60% RH, no significant corrosion damage occurred, despite the very high concentration of SO_2 in the atmosphere.

Another study, performed in Spain, indicates that humidity played a more important role than levels of atmospheric contaminants in predicting corrosion of painted steel [39]. However, without quantitative data of pollutant levels for Madrid and Hospitalet, it is impossible to rule out a combination of humidity and airborne pollutants as the major factor in determining coating performance. In this study, 60 μm chlorinated rubber was applied to clean steel. Painted samples and coupons of bare steel and zinc were exposed in dry rural, dry urban, humid industrial, and humid coastal areas. The results after two years are given in Table 6.3.

TABLE 6.3
Performance of Bare Steel and Coated Panels

Location	Type of atmosphere	Humid/Dry	Corrosion of bare steel (μm/year)	Degree of oxidation of painted surface (%) after 2 years
El Pardo	Rural	Dry	14.7	0
Madrid	Urban	Dry	27.9	0
Hospitalet	Industrial	Humid	52.7	0.3
Vigo	Coastal	Humid	62.6	16

Modified from: Morcillo, M. and S. Feliu, *Proc., Corrosio i Medi Ambient,* Universitat de Barcelona, Barcelona, 1986, 312.

TABLE 6.4
Comparison of Bare Steel and Painted Panels at Sheffield and Calshot

	Sheffield	Calshot
Type of environment	Industrial	Marine
Rate of corrosion of mild steel over 5 years, μm/year	109	28
Life to failure of a 4-coat painting scheme, years	6.1	6.0

Modified from: Sixth Report of the Corrosion Committee, Special Report No. 66, Iron and Steel Institute, London, 1959.

In a 1950s British study of bare steel and painted panels carried out at Sheffield, an industrial site that had heavy atmospheric pollution at the time, and Calshot, a marine site, the same overall result — no correlation between airborne chemicals and corrosion of painted metal — was seen [40]. The corrosion rate of bare steel in Sheffield was about four times that at Calshot, but the performance of the painted panels was about the same (see Table 6.4).

The same trend was seen in a Portuguese study of zinc-rich coatings exposed at Sines (marine atmosphere) and Lavradio (industrial). Bare metal coupons of mild steel corroded nearly twice as much at Lavradio as at Sines, and zinc corroded almost four times as much at Lavradio. The 18 coatings studied at both sites did not reflect that difference [41].

REFERENCES

1. Ranby, B. and Rabek, J.F., *Photodegradation, Photo-oxidation and Photostabilization of Polymers: Principles and Application*, Wiley Interscience, New York, 1975, 242.
2. Forsgren, A. and Appelgren, C., *Performance of Organic Coatings at Various Field Stations After 5 Years' Exposure*, Report 2001:5E, Swedish Corrosion Institute, Stockholm, 2001.
3. Krejcar, E. and Kolar, O., *Prog. Org. Coat.*, 3, 249, 1973.
4. Hare, C.H., *J. Prot. Coat. Linings*, 17, 73, 2000.
5. Berg, C.J., Jarosz, W.R. and Salanthe, G.F., *J. Paint Technol.*, 39, 436, 1967.
6. Nichols, M.E. and Darr, C.A., *J. Coat. Technol.*, 70, 885, 1998.
7. Oosterbroek, M.L. et al., *J. Coat. Technol.*, 63, 55, 1991.
8. Miller, C.D., *J. Amer. Oil Chem. Soc.*, 36, 596, 1959.
9. Marshall, N.J., *Off. Dig.*, 29, 792, 1957.
10. Fitzgerald, E.B., in ASTM Bulletin 207 TP-137, American Society for Testing and Materials, Philadelphia, PA, 1955, 650.
11. Hare, C.H., *J. Prot. Coat. Linings*, 13, 65, 1996.
12. Allen, N.S. et al., *Prog. Org. Coat.*, 32, 9, 1997.

13. Lindqvist, S.A., *Corrosion,* 41, 69, 1985.
14. Funke, W., Zorll, U. and Murthy, B.G.K., *J. Coat. Technol.,* 68, 210, 1996.
15. Huldén, M. and Hansen, C.M., *Prog. Org. Coat.,* 13, 171, 1985.
16. Ferlauto, E.C. et al., *J. Coat. Technol.,* 66, 85, 1994.
17. Hare, C.H., *J. Prot. Coat. Linings,* 13, 59, 1996.
18. Lefebvre, D.R. et al., *J. Adhesion Sci. Technol.,* 5, 210, 1991.
19. Perera, D.Y., *Prog. Org. Coat.,* 44, 55, 2002.
20. Perera, D.Y and Vanden Eynde, D., *J. Coat. Technol.,* 59, 55, 1987.
21. Prosser, J.L., *Mod. Paint and Coat.,* 47, July 1977.
22. Sato, K. and Inoue, M., *Shikizai Kyosaishi,* 32, 394, 1959. (Summarized in Hare, C.H., *J. Prot. Coat. Linings,*13, 59, 1996.)
23. Perera, D.Y and Vanden Eynde, D., in *Proc. Vol.1, XVIth FATIPEC Congress,* Fédération d'Associations de Techniciens des Industries des Peintures, Vernis, Emaux et Encres d'Imprimerie de l'Europe Continentale (FATIPEC), Paris, 1982, 129.
24. Perera, D.Y., *Prog. Org. Coat.,* 28, 21, 1996.
25. Perera, D.Y. and Vanden Eynde, D., in *Proc. XXth FATIPEC Congress,* Fédération d'Associations de Techniciens des Industries des Peintures, Vernis, Emaux et Encres d'Imprimerie de l'Europe Continentale (FATIPEC), Paris, 1990, 125.
26. Perera, D.Y., Stress phenomena in organic coatings, in *Paint and Coatings Testing Manual,* 14th ed. Of Gardner-Sward Handbook, Koleske, J.V., Ed., ASTM, Philadelphia, PA, 1995.
27. Leidheiser, H., Wang, W. and Igetoft, L., *Prog. Org. Coat.,* 11, 19, 1983.
28. Funke, W., *Ind. Eng. Chem. Prod. Res. Dev.,* 24, 343, 1985.
29. Miszczyk, A. and Darowicki, K., *Prog. Org. Coat.,* 46, 49, 2003.
30. Maitland, C.C. and Mayne, J.E.O., *Off. Dig.,* 34, 972, 1962.
31. Cherry, B.W. and Mayne, J.E.O., *Proc. First International Congress on Metallic Corrosion,* Butterworths, London. 1961.
32. Mayne, J.E.O., *Trans. Inst. Met. Finish.,* 41, 121, 1964.
33. Cherry, B.W. and Mayne, J.E.O., *Off. Dig.,* 37, 13, 1965.
34. Mayne, J.E.O., *JOCCA,* 40, 183, 1957.
35. Sampers, J., *Polymer Degradation and Stability,* 76, 455, 2002.
36. Carew, J. A. et al., *Weathering performance of industrial atmospheric coatings systems in the Arabian Gulf,* Proc. Corrosion '94, NACE, Houston, 1994, Paper 445.
37. Rendahl, B., Igetoft, L. and Forsgren, A., *Field testing of anticorrosion paints at sulphate and sulphite mills,* in Proc. 9th International Symposium on Corrosion in the Pulp and Paper Industry, PAPRICAN, Quebec, 1998.
38. Özcan, M., Dehri, I. and Erbil, M., *Prog. Org. Coat.,* 44, 279, 2002.
39. Morcillo, M. and Feliu, S., *Quantitative data on the effect of atmospheric contamination in coatings performance,* Proc. Corrosio i Medi Ambient, Universitat de Barcelona, Barcelona, 1986, 312.
40. *Sixth Report of the Corrosion Committee,* Special Report No. 66, Iron and Steel Institute, London, 1959.
41. Almeida, E.M., Pereira, D. and Ferreira, M.G.S., An electrochemical and exposure study of zinc rich coatings, in *Proc. Advances in Corrosion Protection by Organic Coatings* (Vol. 89-13), Scantlebury, D. and Kendig, M., Eds., The Electrochemical Society Inc., Pennington, 1989, 486.

7 Corrosion Testing — Background and Theoretical Considerations

The previous chapter described the aging process of an organic coating, which leads to coating failure. The major factors that cause aging and degradation of organic coatings are UV radiation, moisture, heat, and chemical damage. Unfortunately for coating formulators, buyers, and researchers, aging and breakdown of a good coating on a well-prepared substrate takes several years to happen in the field. Knowledge about the suitability of a particular coating is, of course, required on a much shorter time span (usually "right now"); decisions about reformulating, recommending, purchasing, or applying a paint can often wait for a number of weeks or even a few months while test data is collected. Years, however, are out of the question. This explains the need for accelerated testing methods. The purpose of accelerated testing is to duplicate in the laboratory, as closely as possible, the aging of a coating in outdoor environments — but in a much shorter time.

This chapter considers testing the corrosion-protection ability of coatings used in atmospheric exposure. The term "atmospheric exposure" is understood to include both inland and coastal climates, with atmospheres ranging from industrial to rural. Tests used for underwater or offshore applications are not within the scope of this chapter. A very brief explanation of some commonly used terms in corrosion testing of coatings is provided at the end of this chapter.

7.1 THE GOAL OF ACCELERATED TESTING

The goal of testing the corrosion-protection ability of a coating is really to answer two separate questions:

1. Can the coating provide adequate corrosion protection?
2. Will the coating continue to provide corrosion protection over a long period?

The first question is simple: Is the coating any good at preventing corrosion? Does it have the barrier properties, or the inhibitive pigments, or the sacrificial pigments to ensure that the underlying metal does not corrode? The second question is how will the coating hold up over time? Will it rapidly degrade and become useless? Or will it show resistance to the aging processes and provide corrosion protection for many years?

The difference may seem unimportant; however, there are advantages to separating the two questions. Testing a coating for initial corrosion protection is relatively inexpensive and straightforward. The stresses — water, heat, electrolyte — that cause corrosion of the underlying metal are exaggerated and then the metal under the coating is observed for corrosion. However, trying to replicate the aging process of a coating is expensive and difficult for several reasons:

1. Coatings of differing type cannot be expected to have a similar response to an accentuated stress.
2. Scaling down wet-dry cycles changes mass transport phenomena.
3. Climate variability means that the balance of stresses, and subsequent aging, is different from site to site.

7.2 WHAT FACTORS SHOULD BE ACCELERATED?

The major weathering stresses that cause degradation of organic coatings are:

- UV radiation
- Water and moisture
- Temperature
- Ions (salts such as sodium chloride and calcium chloride) and chemicals

The first of these weathering factors is unique to organic coatings; the latter three are also major causes of corrosion of bare metals. Most testing tries to reproduce natural weathering and accelerate it by accentuating these stresses. However, it is critically important to not overaccentuate them. To accelerate corrosion, we scale **up** temperature, salt loads, and frequency of wet-dry transitions; therefore, we must scale **down** the duration of each temperature–humidity step. The balances of mass transport phenomena, electrochemical processes, and the like necessarily change with every accentuation of a stress. The more we scale, the more we change the balances of transport and chemical processes from that seen in the field and the farther we step from real service performance. The more we force corrosion in the laboratory, the less able are we to accurately predict field performance.

For example, a common method of increasing the rate of corrosion testing is to increase the temperature. For certain coatings, the transport of water and oxygen increase markedly at elevated temperature. Even a relatively small increase in temperature above the service range results in large changes in these coating properties. Such coatings are especially sensitive to artificially elevated temperatures in accelerated testing, which may never be seen in service. Other coatings, however, do not see strongly increased oxygen and water transport at the same elevated temperature. An accelerated test at elevated temperatures of these two coatings may falsely show that one was inferior to the other, when in reality both give excellent service for the intended application.

And, of course, interactions between stresses are to be expected. Some major interactions that the coatings tester should be aware of include:

- **Frequency of temperature/humidity cycling.** Because the corrosion reaction depends on supplies of oxygen and water, the accelerated test must correctly mimic the mass transport phenomena that occur in the field. There is a limit to how much we can scale down the duration of a temperature–humidity cycle in order to fit more cycles in a 24-hour period. Beyond that limit, the mass transport occurring in the test no longer mirrors that seen in the field.

- **Temperature/salt load/relative humidity (RH).** The balance of these factors helps to determine the size of the active corrosion cell. If that is not to scale in the accelerated test, the results can diverge greatly from that seen in actual field service. Ström and Ström [1] have described instances of this imbalance in which high salt loads combined with low temperatures led to an off-scale cell.
- **Type of pollutant/RH.** Salts such as sodium chloride (NaCl) and calcium chloride ($CaCl_2$) are hygroscopic but liquefy at different RHs. NaCl liquefies at 76% RH and $CaCl_2$ at 35% to 40% RH (depending on temperature). At an intermediate RH, for example 50% RH, the type of salt used can determine whether or not a thin film of moisture forms on the sample surface due to hygroscopic salts.

Various polymers, and therefore coating types, react differently to a change in one or more of these weathering stresses. Therefore, in order to predict the service life of a coating in a particular application, it is necessary to know not only the environment — average time of wetness, amounts of airborne contaminants, UV exposure, and so on — but also how these weathering stresses affect the particular polymer [2].

7.2.1 UV EXPOSURE

UV exposure is extremely important in the aging and degradation of organic coatings. As the polymeric backbone of a coating is slowly broken down by UV light, the coating's barrier properties can be expected to worsen. However, UV exposure's importance in anticorrosion paints is strictly limited. This is because a coating can be protected from UV exposure simply by painting over it with another paint that does not transmit light.

The role of UV exposure in testing anticorrosion paints may be said to be "pass/fail." Knowing if the anticorrosion paint is sensitive to UV light is important. If it is, then it will be necessary to cover the paint with another coating to protect it from the UV light. This additional coating is routinely done in practice because the most important class of anticorrosion paints, epoxies, are notoriously sensitive to UV stress. It does not prevent epoxies from providing excellent service; rather, it merely protects them from the UV light.

Because UV light itself plays no role in the corrosion process, the need for UV stress in an accelerated corrosion test is questionable.

7.2.2 MOISTURE

There are as many opinions about the proper amount of moisture to use in accelerated corrosion testing of paints as there are scientists in this field. The reason is almost certainly because the amount and form of moisture varies drastically from site to site. The global atmosphere, unless it is locally polluted (e.g., by volcanic activity or industrial facilities), is made up of the same gases everywhere: nitrogen, oxygen, carbon dioxide, and water vapor. Nitrogen and carbon dioxide do not affect coated metal. Oxygen and water vapor, however, cause aging of the coating and corrosion of the underlying metal. The amount of oxygen is more or less constant everywhere, but the amount of water vapor in the air is not. It varies depending on location, time of day, and season [3].

The form of water also varies: water vapor in the atmosphere is a gas, and rain or condensation is a liquid. To further complicate things, water in the coating can go from one form to another; whether or not this happens — and how fast —depends on both the temperature and the RH of the air.

It is often noted that water vapor may have more effect on the coating than does liquid water. For nonporous materials, there is no theoretical difference between permeation of liquid water and that of water vapor [4]. Coatings, of course, are not solid, but rather contain a good deal of empty space, for example:

1. Pinholes are created during cure by escaping solvents.
2. Void spaces are created by crosslinking. As crosslinking occurs during cure, the polymer particles cease to move freely. The increasing restrictions on movement mean that the polymer molecules cannot be "packed" efficiently in the shrinking film. Voids are created as solvent evaporates from the immobilized polymer matrix.
3. Void spaces are created when polymer molecules bond to a substrate. Before a paint is applied, polymer molecules are randomly disposed in the solvent. Once applied to the substrate, polar groups on the polymer molecule bond at reactive sites on the metal. Each bond created means reduced freedom of movement for the remaining polymer molecules. As more polar groups bond on reactive sites on the metal, the polymer chain segments between bonds loop upward above the surface (see Figure 7.1). The looped segments occupy more volume and form voids at the surface, where water molecules can aggregate [5].
4. Spaces form between the binder and the pigment particles. Even under the best circumstances, areas arise on the surface of the pigment particle where

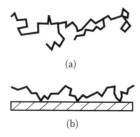

(a)

(b)

FIGURE 7.1 Looped polymer segments above the metal surface.

the binder and the particle may be in extremely close physical proximity but are not chemically bonded. This area between binder and pigment can be a potential route for water molecules to slip through the cured film.

Ström and Ström [1] have offered a definition of wetness that may be useful in weighing vapor versus liquid water. They have pointed out that NaCl liquidates at 76% RH, and CaCl$_2$ liquidates at 35% to 40% RH (depending on temperature). NaCl is by far the most commonly used salt in corrosion testing. It seems reasonable to assume that, unless the electrolyte spray/immersion/mist step in an accelerated test is followed by a rinse, a hygroscopic salt residue will exist on the sample surface. At conditions below condensing but above the liquidation point for NaCl, the hygroscopic residue can give rise to a thin film of moisture on the surface. Therefore, conditions at 76% RH or more should be regarded as wet. Time of wetness (TOW) for any test would thus be the amount of time in the cycle where the RN is at 76% or higher.

7.2.3 DRYING

A critical factor in accelerated testing is drying. Although commonly ignored, drying is as important as moisture. The temptation is to make the corrosion go faster by having as much wet time as possible (i.e., 100% wet). However, this approach poses two problems:

1. Studies indicate that corrosion progresses most rapidly during the transition period from wet to dry [6–10].
2. The corrosion mechanism of zinc in 100% wet conditions is different from that usually seen in actual service.

7.2.3.1 Faster Corrosion during the Wet–Dry Transition

Stratmann and colleagues have shown that 80% to 90% of atmospheric corrosion of iron occurs at the end of the drying cycle [7]; similar studies exist for carbon steel and zinc-coated steel. Ström and Ström [1] have reported that the effect of drying may be even more pronounced on zinc than on steel. Ito and colleagues [6] have provided convincing data of this as well. In their experiments, the drying time ratio, R_{dry}, was defined as the percentage of the time in each cycle during which the sample is subjected to low RH:

$$R_{dry} = \frac{T_{drying}}{T_{cycle}} \bullet 100\%$$

The drying condition was defined as 35°C and 60% RH; the wet condition was defined as 35°C and constant 5% NaCl spray (i.e., salt spray conditions). T_{cycle} is the total time, wet plus dry, of one cycle, and T_{drying} is the amount of time at 60% RH, 35°C during one cycle. Cold-rolled steel and galvanized steels with three zinc-coating thicknesses were tested at $R_{dry} = 0$, 50, and 93.8%. For all four substrates, the highest amount of steel weight loss was seen at $R_{dry} = 50\%$.

In summary, corrosion on both steel and zinc-coated steel substrates is slower if no drying occurs. This finding seems reasonable because, as the electrolyte layer becomes thinner while drying, the amount of oxygen transported to the metal surface increases, enabling more active corrosion [11, 12]. A similar highly active phase can be expected to occur during rewetting under cyclic conditions.

Readers interested in a deeper understanding of this process may find the works of Suga [13] and Boocock [14] particularly helpful.

7.2.3.2　Zinc Corrosion — Atmospheric Exposure vs. Wet Conditions

A drying cycle is an absolute must if zinc is involved either as pigment or as a coating on the substrate. The corrosion mechanism that zinc undergoes in constant humidity is quite different from that observed when there is a drying period. In field service, alternating wet and dry periods is the rule. Under these conditions, zinc can offer extremely good real-life corrosion protection — but this would never be seen in the laboratory if only constant wetness is used in the accelerated testing. This apparent contradiction is worth exploring in some depth.

Although this is a book about paints, not metallic corrosion, it becomes necessary at this point to devote some attention to the corrosion mechanisms of zinc in dry versus wet conditions. The reason for this is simple: zinc-coated steel is an important material for corrosion prevention, and it is frequently painted. Accelerated tests are therefore used on painted, zinc-coated steel. In order to obtain any useful information from accelerated testing, it is necessary to understand the chemistry of zinc in dry and wet conditions.

In normal atmospheric conditions, zinc reacts with oxygen to form a thin oxide layer. This oxide layer in turn reacts with water in the air to form zinc hydroxide ($Zn[OH]_2$), which in turn reacts with carbon dioxide in air to form a layer of basic zinc carbonate [15-17]. Zinc carbonate serves as a passive layer, effectively protecting the zinc underneath from further reaction with water and reducing the amount of corrosion.

When zinc-coated steel is painted and then scribed to the steel, the galvanic properties of the zinc-steel system determine whether, and how much, corrosion will take place under the coating. Two mechanisms cause the growth of red rust and undercutting from the scribe [1, 6, 18-21]:

1. The first reaction is a galvanic cell located at the scribe. The anode is the metal exposed in the scribe, and the cathode is the adjacent zinc layer under the paint.
2. The second reaction is located not at the scribe but rather at the leading edge of the zinc corrosion front. Anodic dissolution of zinc occurs from the top of the zinc layer and works downward to the steel.

Ito and colleagues have postulated that the magnitudes and the comparative ratio of these two mechanisms changes with the amount of water available. When they repeated their experiments with R_{dry} on painted, cold-rolled and galvanized steels,

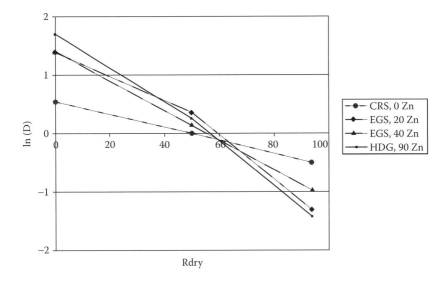

FIGURE 7.2 Natural log of underfilm corrosion, as a function of drying ratio for cold-rolled steel, electrogalvanized (20 g/m² Zn and 40 g/m² Zn), and hot-dipped galvanized (90 g/m² Zn). *Data from:* Ito, Y., Hayashi, K., and Miyoshi, Y., *Iron Steel J.*, 77, 280, 1991.

an interesting pattern emerged. Instead of measuring weight of metal lost, they measured the distance of underfilm corrosion from the scribe. In Figure 7.2, the natural logarithm of the length of underfilm corrosion D, measured by Ito and colleagues, is plotted against the R_{dry} for each of the four coating weights. The relationship between zinc coating thickness, drying ratio, and underfilm corrosion distance is fairly distinct when presented thus.

Ito and colleagues have also proposed that under wet conditions, (i.e., low R_{dry}), more underfilm corrosion is seen on zinc-coated steel than on cold-rolled steel because the following two reactions at the boundary between paint and zinc layer dominate the corrosion:

1. Zinc dissolves anodically at the front end of corrosion.
2. In the blister area behind the front end of corrosion, zinc at the top of the zinc layer dissolves due to OH, which is generated by cathodic reaction.

However, if conditions include high R_{dry}, then underfilm corrosion is less on galvanized steel than on cold-rolled steel, for the following reasons:

1. The total supply of water and chloride (Cl⁻) is reduced, limiting cell size at front end and zinc anodic dissolution area.
2. The electrochemical cell at the scribe is reduced.
3. Zinc is isolated from the wet corrosive environment fairly early. A protective film can form on zinc in dry atmosphere. The rate of zinc corrosion is suppressed in further cycling.
4. The zinc anodic dissolution rate is reduced because the Cl⁻ concentration at the front end is suppressed.

It should be noted that the 90 g/m^2 zinc coating in this study is hot-dipped galvanized, and the two thinner coatings are electrogalvanized. It may be that differences other than zinc thickness — for example, structure and morphology of the zinc coating — play a not yet understood role. Further research is needed in this area, to understand the role played by zinc layer structure and morphology in under–cutting.

7.2.3.3 Differences in Absorption and Desorption Rates

The rate at which a coating absorbs water is not necessarily the same as the rate at which it dries out. Some coatings have nearly the same absorption and desorption rates, whereas others show slower drying than wetting, or vice versa.

In constant stress testing, in which samples are always wet or always dry, this difference does not become a factor. However, as soon as wet-dry cycles are introduced, the implications of a difference between absorption and desorption rates becomes highly important. Two coatings with roughly similar absorption rates can have vastly different desorption rates. The duration of wet and dry periods in modern accelerated tests is measured in hours, not days, and it is quite possible that, for a coating with a slower desorption rate, the drying time in each cycle is shorter than the time needed by the coating for complete desorption. In such cases, the coating that desorbs more slowly than it absorbs can accumulate water.

The problem is not academic. Lindqvist [22] has studied absorption and desorption rates for epoxy, chlorinated rubber, linseed oil, and alkyd binders, using a cycle of 6 hours of wet followed by 6 hours of drying. An epoxy coating took up 100% of its possible water content in the wet periods but never dried out in the drying periods. Conversely, a linseed oil coating in this study never reached its full saturation during the 6-hour wet periods but dried out completely during the drying periods.

Lindqvist has pointed out that the difference in the absorption and desorption rates of a single paint, or of different types of paint, could go far in explaining why cyclic accelerated tests often do not produce the same ranking of coatings as does field exposure. There is a certain risk to subjecting different types of coatings with unknown absorption and desorption characteristics to a cyclic wet-dry accelerated regime. The risk is that the accelerated test will produce a different ranking from that seen in reality. It could perhaps be reduced by some preliminary measurements of water uptake and desorption; an accelerated test can then be chosen with both wet times and drying times long enough to let all the paints completely absorb and desorb.

7.2.4 TEMPERATURE

Temperature is a crucial variable in any accelerated corrosion testing. Higher temperature means more energy available, and thus faster rates, for the chemical processes that cause both corrosion and degradation of cured films. Increasing the temperature — within limits — does not alter the corrosion reaction at the metal surface; it merely speeds it up. A potential problem, however, is what the higher temperature does to the binder. If the chemical processes that cause aging of the binder were simply speeded up without being altered, elevated temperature would pose no problem. But this is not always the case.

Every coating is formulated to maintain a stable film over a certain temperature range. If that range is exceeded, the coating can undergo transformations that would not occur under natural conditions [3]. The glass transition temperature (T_g) of the polymer naturally limits the amount of acceleration that can be forced by increasing heat stress. Testing in the vicinity of the T_g changes the properties of the coatings too much, so that the paint being tested is not very much like the paint that will be used in the field — even if it came from the same can of paint.

7.2.5 CHEMICAL STRESS

When the term "chemical stress" is used in accelerated testing, it usually means chloride-containing salts in solution, because airborne contaminants are believed to play a very minor role in paint aging. See Chapter 6 for information about air bourne contaminents.

Testers may be tempted to force quicker corrosion testing by increasing the amount of chemical stress. Steel that corrodes in a 0.05% sodium chloride (NaCl) solution will corrode even more quickly in 5% NaCl solution; the same is true for zinc-coated steel. The problem is that the amount of acceleration is different for the two metals. An increase in NaCl content has a much more marked effect for zinc-coated substrates than for carbon steel substrates. Ström and Ström [1] have demonstrated this effect in a test of weakly accelerated outdoor exposure of painted zinc-coated and carbon steel samples. In this weakly accelerated test, commonly known as the "Volvo Scab" test, samples are exposed outdoors and sprayed twice a week with a salt solution. Table 7.1 gives the results after 1 year of this test, using different levels of NaCl for the twice-weekly spray.

TABLE 7.1
Average Creep from Scribe after 1 Year Weakly Accelerated Field Exposure

Material	Outdoor samples sprayed twice per week with:		
	0.5% NaCl	1.5% NaCl	5% NaCl
Mean for all electrogalvanized and hot-dipped galvanized painted samples	1.3 mm	2.0 mm	3.1 mm
Mean for all cold-rolled steel painted samples	6.2 mm	8.2 mm	9.6 mm

Modified from: Ström, M. and Ström, G., *SAE Technical Paper Series, 932338,* Society of Automotive Engineers, Warrendale, Pennsylvania, 1993.

From this study, it can be seen that raising the chloride load has a much stronger effect on painted zinc-coated substrates than on painted carbon-steel substrates. It is known that for bare metals, the zinc corrosion rate is more directly dependent than the carbon steel corrosion rate on the amount of pollutant (NaCl in this case). This relationship may be the cause of the results in the table above. In addition, higher salt levels leave a heavier hygroscopic residue on the samples (see Section 7.2.3); this may have caused a thicker moisture film at RH levels above 76%.

Boocock [23] reports another problem with high NaCl levels in accelerated tests: high saponification reactions, which are not seen in the actual service, can occur at high NaCl loads. Coatings that give good service in actual field exposures can wrongly fail an accelerated test with a 5% NaCl load.

Increasing the level of NaCl increases the rate of corrosion of painted samples, but the amount of acceleration is not the same for different substrates. As the NaCl load is increased, the range of substrates or coatings that can be compared with each other in the test must narrow. A low salt load is recommended for maximum reliability.

Another approach is to reduce the frequency of salt stress. Most cyclic tests call for salt stress between 2 and 7 times per week. Smith [24], however, has developed a cyclic test for the automotive industry that uses 5-minute immersion in 5% NaCl once every 2 weeks. The high salt load — typical for when the test was developed — is offset by the low frequency.

How much salt is too much? There is no consensus about this, but several agree that the 5% NaCl used in the famous salt spray test is too high for painted samples. Some workers suggest that 1% NaCl should be a natural limit. Some of the suggested electrolyte solutions at lower salt loads (using water as solvent) are:

0.05% (wt) NaCl and 0.35% ammonium sulfate, $(NH_4)_2SO_4$ [25]

0.5% NaCl + 0.1% $CaCl_2$ + 0.075% $NaHCO_3$ [26]

0.9% NaCl + 0.1% $CaCl_2$ + 0.25% $NaHCO_3$ [27]

7.2.6 ABRASION AND OTHER MECHANICAL STRESSES

While in service, coatings undergo external mechanical stresses, such as:

• Abrasion (also called *sliding wear*)
• Fretting wear
• Scratching wear
• Flexing
• Impingement or impact

These stresses are not of major importance in corrosion testing. Even though some damage to the coating is usually needed to start corrosion, such as a scribe down to the metal, the mechanical damage in and of itself does not cause corrosion. This is not to say that the area is unimportant: a feature of good anticorrosion coatings is that they can contain the amount of corrosion by not allowing undercutting to spread far from the original point of damage. Mechanical stress may be viewed in a manner similar to that for UV exposure: depending on the service application, it can be a "pass/fail" type of test. For example, in applications that will be exposed to a lot of stone chipping (i.e., because of proximity to a highway), impact testing may be needed. If the anticorrosion coating fails the impact test, then a covering coat tailored to this requirement may be needed.

There are several excellent reviews of external mechanical stresses, including details of their causes, their effects on various coating types, and the test methods used to measure a coating's resistance to them. For more information, the reader is directed toward several existing publications [28-30].

7.2.7 IMPLICATIONS FOR ACCELERATED TESTING

Traditionally, accelerated testing of organic coatings has been attempted in the laboratory by exaggerating the stresses (heat, moisture, UV, and salt exposure) that age the coating. The prevailing philosophy has been that more stress = more acceleration.

The previous sections have discussed why this prevailing philosophy is flawed. In this section, some limitations on stresses are proposed:

- Temperatures cannot be elevated above or anywhere near the T_g of the polymer.
- Moisture is important, but a drying cycle is equally important.
- Salt levels should be lower than those commonly used today.
- UV exposure is probably not necessary.

7.3 WHY THERE IS NO SINGLE PERFECT TEST

A great deal of research has gone into understanding the aging process of coatings, and attempts to replicate it more accurately and quickly in laboratories. Great advances have been made in the field, and even more advances are expected in the future. Still, we will never see one perfect accelerated test that can be used to predict coating performance anywhere in the world, on all coating types and all substrates.

There are several reasons why not:

- Different sites around the world have different climates, stresses, and aging mechanisms.
- Different coatings have different weaknesses, and will not respond identically to an accentuated stress in the laboratory.
- It is not possible to accentuate all weathering factors, and still maintain the balance between them that exists in the field.

These are discussed in more detail in the following sections.

7.3.1 DIFFERENT SITES INDUCE DIFFERENT AGING MECHANISMS

Sites can differ dramatically in weather. Take, for example, a bridge connecting Prince Edward Island to the Canadian mainland and a bridge connecting the island of Öland to the Swedish mainland. At first glance, one could say that these two sites are roughly comparable. Both are bridges standing in the sea, located closer to the North Pole than to the equator. Yet, these two sites induce different stresses in paints. A coating used on the first bridge would undergo much higher mechanical stress, due to heavy floes of sea ice. It would also see much higher salt loads because the Atlantic Ocean has a higher salt concentration than does the Baltic Sea. If these two sites, which at first glance seemed similar, can induce some differences in aging mechanisms, then the difference must be even more drastic between such coastal sites as Sydney, Vladivostok, and Rotterdam or between inland sites such as Aix-en-Provence, Brasilia, and Cincinnati.

The point is not academic; it is crucially important for choosing accelerated tests. A mechanically tough coating that is not particularly susceptible to salt would perform well at both sites, but an equally mechanically tough coating that allows some slight chloride permeation may fail at Prince Edward Island and succeed at Öland.

A study of coated panels exposed throughout pulp and paper mills in Sweden by Rendahl and Forsgren [31] illustrates the classic problem of using accelerated tests to predict coating performance: the ranking of identical samples can change from site to site. In this study, 23 coating-substrate combinations were exposed at 12 sites in two pulp and paper mills for 5 years. The sites with the most corrosion were the roofs of a digester house and a bleach plant at the sulphate mill. Although these two locations had similar characteristics — same temperature, humidity, and UV exposure — they produced different rankings of coated samples. Both locations agree on the worst sample, but little else. An alkyd paint that gave good results on the bleach plant roof had abysmally poor results on the other roof. Conversely, an acrylic that had significant undercutting on the bleach plant roof performed well on the digester house roof.

These results illustrate why there is no "magic bullet": an accelerated test that correctly predicts the ranking of the 23 samples at the digester house roof may be wildly wrong in predicting the ranking of the same samples at the bleach plant roof of the same mill.

Glueckert [32] has reported the same phenomenon based on a study of gloss loss of six coating systems exposed at both Colton, California, and East Chicago, Indiana. The East Chicago location had an inland climate, with a temperature range of $-23°C$ to $38°C$. The Colton site had higher temperature, more intense sunlight, and blowing sand. The loss of gloss and ranking of the six coatings is shown in Table 7.2. The two sites identified the same best and worst coating, but ranked the four in between differently.

Another study of coatings exposed at various field stations throughout Sweden [2] found no correlations between sites in the corrosion performances of the identical samples, either in the amount of corrosion or in the ranking at each site. In this study, identifying a coating as "always best" or "always worst" was not possible.

TABLE 7.2
Exposure Results from Colton, California, and East Chicago, Indiana

Coating	Gloss loss (%) E. Chicago	Gloss loss (%) Colton	Ranking, E. Chicago	Ranking, Colton
Epoxy-urethane	3	0	1	1
Urethane	38	31	2	3
Waterborne alkyd	56	6	3	2
Epoxy B	65	83	4	5
Acrylic alkyd	68	77	5	4
Epoxy A	98	98	6	6

Data from: Glueckert, A.J., Correlation of accelerated test to outdoor exposure for railcar exterior coatings, in *Proc. Corros. 94*, NACE, Houston, 1994, Paper 596.

Even if only one coating and one substrate were to be tested, it would not be possible to design an accelerated test that would perfectly suit all the exposure sites mentioned in this section — much less all the sites in the world.

7.3.2 DIFFERENT COATINGS HAVE DIFFERENT WEAKNESSES

Cured coatings are commonly thought of as simple structures: the usual depiction is a layer of binder containing pigment particles. The general view is that of a homogenous, continuous, solid binder film reinforced with pigment particles. In reality, the cured coating is a much more complex structure.

For one thing, instead of being a solid, it contains lots of empty space: pinholes, voids after crosslinking, gaps between pigment and binder, and so on. All of these voids are potential routes for water molecules to slip through the cured film. What is important for accelerated testing is that the amount of empty space in the coating is not constant — it can change during weathering, as both the binder and the pigment change. Some pigments, such as passivating pigments, are slowly consumed, causing the empty space between pigment and binder to increase. Other pigments immediately corrode on their surface. The increased volume of the corrosion products can decrease the empty space between particles and binder.

Binders also change with time, for many reasons. The stresses in the binder caused by film formation can be increased, or relieved, during aging. The magnitude of the stresses caused by film formation, and what happens to these stresses upon weathering, depends to a large extent on the type of polymer used for the binder. The same could be said for UV degradation, or any stress that ages binders: the binder's reaction, both in mechanism and in magnitude, depends to a large extent on the specific polymer used. Even if only one exposure site were really to be used, it would not be possible to design an accelerated test that would be suitable for all binders and pigments.

7.3.3 STRESSING THE ACHILLES' HEEL

Every coating has its own Achilles' heel — that is, a point of weakness. The ideal test would accelerate all stresses to the same extent. It would then be possible to compare coatings with different aging mechanisms — different Achilles' heels — to each other.

Unfortunately, it is not possible to accentuate all stresses evenly. Furthermore, it is not possible to accentuate all weathering factors and still maintain the balance between them that exists in the field. When we increase the percentage of time with UV load, for example, we change the ratio of light and dark and move a step away from the real diurnal cycle seen in the field.

Because it is not possible to evenly accelerate all aging factors, the best testing tries to imitate an expected failure mechanism. Each test accentuates one or a few stresses that are rate-controlling for a mechanism. By choosing the right test, it is possible to thus probe for certain expected weaknesses in the coating/substrate system. The trick, of course, is to correctly estimate the failure mechanism for a particular application, and thus pick the most suitable test.

REFERENCES

1. Ström, M. and Ström, G., *SAE Technical Paper Series, 932338*, Society of Automotive Engineers, Warrendale, PA, 1993.
2. Forsgren, A. and Appelgren, C., *Performance of organic coatings at various field stations after 5 years' exposure,* SCI Rapport 2001:5E, Swedish Corrosion Institute, Stockholm, 2001.
3. Appelman, B., *J. Coat. Technol.,* 62, 57, 1990.
4. Huldén, M. and Hansen, C.M. *Prog. Org. Coat.,* 13, 171, 1985.
5. Kumins, C.A. et al., *Prog. Org. Coat.,* 28, 17, 1996.
6. Ito, Y., Hayashi, K. and Miyoshi, Y., *Iron Steel J.,* 77, 280, 1991.
7. Stratmann, M., Bohnenkamp, K. and Ramchandran, T., *Corros. Sci.,* 27, 905, 1987.
8. Miyoshi, Y. et al., *SAE Technical Paper Series, 820334*, Society of Automotive Engineers, Warrendale, PA, 1982.
9. Nakgawa, T., Hakuri, H. and Sato, H., *Mater. Process,* 1, 1653, 1988.
10. Brady, R. et al., *SAE Technical Paper Series, 892567*, Society of Automotive Engineers, Warrendale, PA, 1989.
11. Mansfield, F., Atmospheric corrosion, in *Encyclopedia of Materials Science and Engineering,* Vol. 1, Pergamon Press, Oxford, 1986, 233.
12. Boelen, B. et al., *Corros. Sci.,* 34, 1923, 1993.
13. Suga, S., *Prod. Finish.,* 40, 26, 1987.
14. Boocock, S.K., *JPCL,* 11, 64, 1994.
15. Seré, P.R. et al., *J. Scanning Microsc.,* 19, 244, 1997.
16. Odnevall, I. and Leygraf, C., Atmospheric corrosion, in *ASTM STP 1239,* Kirk, W.W. and Lawson, H.H., Eds., American Society for Testing and Materials, Philadelphia, 1994, 215.
17. Almeida, E.M., Pereira, D. and Ferreira, M.G.S., An electrochemical and exposure study of zinc rich coatings, in *Proc. Advances in Corrosion Protection by Organic Coatings,* Scantlebury, D. and Kendig, M., Eds., The Electrochemical Society Inc., Pennington, 1989, 486.

18. Lambert, M.R. et al., *Ind. Eng. Chem. Prod. Res. Dev.*, 24, 378, 1985.

19. Jordan, D.L., Galvanic interactions between corrosion products and their bare metal precursors: A contribution to the theory of underfilm corrosion, in *Proc. Advances in Corrosion Protection by Organic Coatings*, Scantlebury, D. and Kendig, M., Eds., The Electrochemical Society Inc., Pennington, 1989, 30.

20. Jordan, D.L., Influence of iron corrosion products on the underfilm corrosion of painted steel and galvanized steel, in *Zinc-Based Steel Coating Systems: Metallurgy and Performance*, Krauss, G. and Matlock, D.K., Eds., The Minerals, Metals, & Materials Society, Warrendale, PA, 1990, 195.

21. Jordan, D.L, Franks, L.L. and Kallend, J.S., Measurement of underfilm corrosion propagation by use of spotface paint damage, in *Proc. Corrosion '95*, NACE, Houston, 1995, Paper 384.

22. Lindqvist, S.A., *Corrosion*, 41, 69, 1985.

23. Boocock, S.K., Some results from new accelerated testing of coatings, in *Proc. Corrosion '92*, NACE, Houston, 1992, Paper 468.

24. Smith, A.G., *Polym. Mater. Sci. Eng.*, 58, 417, 1988.

25. Mallon, K. et al., Accelerated test program utilizing a cyclical test method and analysis of methods to correlate with field testing, in *Proc. Corrosion '92*, NACE, Houston, 1992, Paper 331.

26. Townsend, H., Development of an improved laboratory corrosion test by the automotive and steel industries, in *Proc. of the 4th Annual ESD Advanced Coatings Conference*, Engineering Society of Detroit, Detroit, MI, 1994.

27. Yau, Y.-H., Hinnerschietz, S.J. and Fountoulakis, S.G., Performance of organic/metallic composite coated sheet steels in accelerated cyclic corrosion tests, in *Proc. Corrosion '95*, NACE, Houston, 1995, Paper 396.

28. Hare, C.H., *J. Prot. Coat. Linings*, 14, 67, 1997.

29. Koleske, J.V., *Paint and Coating Manual: 14th Edition of the Gardner-Sward Handbook.* ASTM, Philadelphia, 1995.

30. *Surface Coatings: Science & Technology.* 2nd ed., Paul, S., Ed., John Wiley & Sons, Chichester, 1996.

31. Rendahl, B. and Forsgren, A., *Field Testing of Anticorrosion Paints at Sulphate and Sulphite Mills*, Report 1997:6E, Swedish Corrosion Institute, Stockholm, 1998.

32. Glueckert, A.J., Correlation of accelerated test to outdoor exposure for railcar exterior coatings, in *Proc. Corros. 94*, NACE, Houston, 1994, Paper 596.

8 Corrosion Testing — Practice

Corrosion tests for organic coatings can be divided into two categories:

1. **Test regimes that age the coating.** These are the accelerated test methods, including single stress tests, such as the salt spray, or cyclic tests such as the American Society for Testing and Materials (ASTM) D5894.
2. **Measurements of coating properties before and after aging.** These tests measure such characteristics as adhesion, gloss, and barrier properties (water uptake).

The aim of the accelerated test regime is to age the coating in a short time in the same manner as would occur over several years' field service. These tests can provide direct evidence of coating failure, including creep from scribe, blistering, and rust intensity. They also are a necessary tool for the measurement of coating properties that can show indirect evidence of coating failure. A substantial decrease in adhesion or significantly increased water uptake, even in the absence of rust-through or undercutting, is an indication of imminent coating failure.

This chapter provides information about:

- Which accelerated tests age coatings
- What to look for after an accelerated test regime is completed
- How the amount of acceleration in a test is calculated, and how the test is correlated to field data
- Why the salt spray test should not be used

8.1 SOME RECOMMENDED ACCELERATED AGING METHODS

Hundreds of test methods are used to accelerate the aging of coatings. Several of them are widely used, such as salt spray and ultraviolet (UV) weathering. A review of all the corrosion tests used for paints, or even the major cyclic tests, is beyond the scope of this chapter. It is also unnecessary because this work has been presented elsewhere; the reviews of Goldie [1], Appleman [2], and Skerry and colleagues [3] are particularly helpful.

The aim of this section is to provide the reader with an overview of a select group of accelerated aging methods that can be used to meet most needs:

- General corrosion tests — all-purpose tests
- Condensation or humidity tests
- Weathering tests (UV exposure)

In addition, some of the tests used in the automotive industry are described. These are tests with proven correlation to field service for car and truck paints, which may, with adaptations, prove useful in heavier protective coatings.

8.1.1 GENERAL CORROSION TESTS

A general accelerated test useful in predicting performance for all types of coatings, in all types of service applications, is the "Holy Grail" of coatings testing. No test is there yet, and none probably ever will be (see Chapter 7). However, some general corrosion tests can still be used to derive useful data about coating performance. The two all-purpose tests recommended here are the ASTM D5894 test and the NORSOK test.

8.1.1.1 ASTM D5894

ASTM D5894, "Standard Practice for Cyclic Salt Fog/UV Exposure of Painted Metal (Alternating Exposures in a Fog/Dry Cabinet and a UV/Condensation Cabinet)," is also called "modified Prohesion" or "Prohesion UV." This test, incidentally, is sometimes mistakenly referred to as "Prohesion testing." However, the Prohesion test does not include a UV stress; it is simply a cyclic salt fog (1 hour salt spray, with 0.35% ammonium sulphate and 0.05% sodium chloride [NaCl], at 23°C, alternating with one drying cycle at 35°C). The confusion no doubt arises because the original developers of ASTM D5894 referred to it as "modified Prohesion."

This test is can be used to investigate both anticorrosion and weathering characteristics. The test's cycle is 2 weeks long and typically runs for 6 cycles (i.e., 12 weeks total). During the first week of each cycle, samples are in a UV/condensation chamber for 4 hours of UV light at 60°C, alternating with 4 hours of condensation at 50°C. During the second week of the cycle, samples are moved to a salt-spray chamber, where they undergo 1 hour of salt spray (0.05% NaCl + 0.35% ammonium sulphate, pH 5.0 to 5.4) at 24°C, alternating with 1 hour of drying at 35°C.

The literature contains warnings about too-rapid corrosion of zinc in this test; therefore, it should not be used for comparing zinc and nonzinc coatings. If zinc and nonzinc coatings must be compared, an alternate (i.e., nonsulphate) electrolyte can be substituted under the guidelines of the standard. This avoids the problems caused by the solubility of zinc sulphate corrosion products. It has also been noted that the ammonium sulphate in the ASTM D5894 electrolyte has a pH of approximately 5; at this pH, zinc reacts at a significantly higher rate than at neutral pH levels. The zinc is unable to form the zinc oxide and carbonates that give it long-term protection.

8.1.1.2 NORSOK

NORSOK is suitable for both corrosion and weathering testing. Its cycle is 168 hours long, and it runs for 25 cycles (i.e., 25 weeks total). Each cycle consists of 72 hours of salt spray, followed by 16 hours drying in air, and then 80 hours of UV condensation (ASTM G53).

The NORSOK test was developed for the offshore oil industry, particularly the conditions found in the North Sea. The test is part of the NORSOK M-501 standard, which provides requirements for materials selection, surface preparation, paint application, inspection, and so on for coatings used on offshore platforms.

8.1.2 CONDENSATION OR HUMIDITY

Many tests are based on constant condensation or humidity. Incidentally, constant condensation is not the same as humidity testing. Condensation rates are higher in the former than the latter because, in constant condensation chambers, the back sides of the panels are at room temperature and the painted side faces water vapor at 40°C. This slight temperature differential leads to higher water condensation on the panel. If no such temperature differential exists, the conditions provide humidity testing in what is known as a "tropical chamber." The Cleveland chamber is one example of condensation testing; a salt spray chamber with the salt fog turned off, the heater turned on, and water in the bottom (to generate vapor) is a humidity test.

Constant condensation or humidity testing can be useful as a test for barrier properties of coatings on less-than-ideal substrates — for example, rusted steel. Any hygroscopic contaminants, such as salts entrapped in the rust, attract water. On new construction, or in the repainting of old construction, where it is possible to blast the steel to Sa2^1/$_2$, these contaminants are not be found. However, for many applications, dry abrasive or wet blasting is not possible, and only handheld tools such as wire brushes can be used. These tools remove loose rust but leave tightly adhering rust in place. And, because corrosion-causing ions, such as chloride (Cl^-), are always at the bottom of corrosion pits, the matrix of tightly adhering rust necessarily contains these hygroscopic contaminants. In such cases, the coating must prevent water from reaching the intact steel. The speed with which blisters develop under the coating in condensation conditions can be an indication of the coating's ability to provide a water barrier and thus protect the steel.

Various standard test methods using constant condensation or humidity testing include the International Organization for Standardizaton (ISO) 6270, ISO 11503, the British BS 3900, the North American ASTM D2247, ASTM D4585, and the German DIN 50017.

8.1.3 WEATHERING

In UV weathering tests, condensation is alternated with UV exposure to study the effect of UV light on organic coatings. The temperature, amount of UV radiation, length (time) of UV radiation, and length (time) of condensation in the chamber are programmable. Examples of UV weathering tests include QUV-A, QUV-B (® Q-Panel Co.), and Xenon tests. Recommended practices for UV weathering are described in the very useful standard ASTM G154 (which replaces the better-known ASTM G53).

8.1.4 CORROSION TESTS FROM THE AUTOMOTIVE INDUSTRY

The automotive industry places great demands on its anticorrosion coatings system and has therefore invested a good deal of effort in developing accelerated tests to help predict the performance of paints in harsh conditions. It should be noted that most automotive tests, including the cyclic corrosion tests, have been developed using coatings relevant to automotive application. These are designed to act quite different from protective coatings. Automotive-derived test methods commonly overlook factors critical to protective coatings, such as weathering and UV factors. In addition, automotive coatings have much lower dry film thickness than do many protective coatings; this is important for mass-transport phenomena.

This section is not intended as an overview of automotive industry tests. Some tests that have good correlation to actual field service for cars and trucks, such as the Ford APGE, Nissan CCT-IV, and GM 9540P [4], are not described here. The three tests described here are those believed to be adaptable to heavy maintenance coatings VDA 621-415, the Volvo Indoor Corrosion Test (VICT), and the Society of Automotive Engineers (SAE) J2334.

8.1.4.1   VDA 621-415

For many years, the automotive industry in Germany has used an accelerated test method for organic coatings called the VDA 621-415 [5]; this test has begun to be used as a test for heavy infrastructure paints also. The test consists of 6 to 12 cycles of neutral salt spray (as per DIN 50021) and 4 cycles in an alternating condensation water climate (as per DIN 50017). The time-of-wetness of the test is very high, which implies poor correlation to actual service for zinc pigments or galvanized steel. It is expected that zinc will undergo a completely different corrosion mechanism in the nearly constant wetness of the test than the mechanism that takes place in actual field service. The ability of the test to predict the actual performance of zinc-coated substrates and zinc-containing paints must be carefully examined because these materials are commonly used in the corrosion engineering field. Also, the start of the test (24 hours of 40°C salt spray) has been criticized as unrealistically harsh for latex coatings.

8.1.4.2   Volvo Indoor Corrosion Test or Volvo-cycle

The VICT [6] was developed — despite its name — to simulate the **outdoor** corrosion environment of a typical automobile. Unlike many accelerated corrosion tests, in which the test procedure is developed empirically, the VICT test is the result of a statistical factorial design [7, 8].

In modern automotive painting, all of the corrosion protection is provided by the inorganic layers and the thin (circa 25 μm) electrocoat paint layer. Protection against UV light and mechanical damage is provided by the subsequent paint layers (of which there are usually three). Testing of the **anticorrosion** or electrocoat paint layer can be restricted to a few parameters, such as corrosion-initiating ions (usually chlorides), time-of-wetness, and temperature. The Volvo test accordingly uses no UV exposure or mechanical stresses; the stresses used are temperature, humidity, and salt solution (sprayed or dipped).

The automotive industry has a huge amount of data for corrosion in various service environments. The VICT has a promising correlation to field data; one criticism that is sometimes brought against this test is that it may tend to produce filiform corrosion at a scribe.

There are four variants of the Volvo-cycle, consisting of either constant temperature together with two levels of humidity or of a constant dew point (i.e., varying temperature and two levels of humidity). The VICT-2 variant, which uses constant temperature and discrete humidity transitions between two humidity levels, is described below.

- **Step I:** 7 hours exposure at 90% relative humidity (RH) and 35°C constant level.
- **Step II:** Continuous and linear change of RH from 90% RH to 45% RH at 35°C during 1.5 hours.
- **Step III:** 2 hours exposure at 45% RH and 35°C constant level.
- **Step IV:** Continuous and linear change of RH from 45% to 90% RH at 35°C during 1.5 hours.

Twice a week, on Mondays and Fridays, step I above is replaced by the following:

- **Step V:** Samples are taken out of the test chamber and submerged in, or sprayed with, 1% (wt.) NaCl solution for 1 hour.
- **Step VI:** Samples are removed from the salt bath; excess liquid is drained off for 5 minutes. The samples are put back into the test chamber at 90% RH so that they are exposed in wetness for at least 7 hours before the drying phase.

Typically the VICT test is run for 12 weeks. This is a good general test when UV is not expected to be of great importance.

8.1.4.3   SAE J2334

The SAE J2334 is the result of a statistically designed experiment using automotive industry substrates and coatings. In the earliest publications about this test, it is also referred to as "PC-4" [4]. The test is based on a 24-hour cycle. Each cycle consists of a 6-hour humidity period at 50°C and 100% RH, followed by a 15-minute salt application, followed by a 17 hours and 45 minute drying stage at 60°C and 50% RH. Typical test duration is 60 cycles; longer cycles have been used for heavier coating weights. The salt concentrations are fairly low, although the solution is relatively complex: 0.5% NaCl + 0.1% $CaCl_2$ + 0.075% $NaHCO_3$.

8.1.5   A TEST TO AVOID: KESTERNICH

In the Kesternich test, samples are exposed to water vapor and sulfur dioxide for 8 hours, followed by 16 hours in which the chamber is open to the ambient environment of the laboratory [2]. This test was designed for bare metals exposed to a polluted industrial environment and is fairly good for this purpose. However, the test's relevance for organically coated metals is highly questionable. For the same reason, the similar test ASTM B-605 is not recommended for painted steel.

8.2 EVALUATION AFTER ACCELERATED AGING

After the accelerated aging, samples should be evaluated for changes. By comparing samples before and after aging, one can find:

- Direct evidence of corrosion
- Signs of coating degradation
- Implicit signs of corrosion or failure

The coatings scientist uses a combination of techniques for detecting macroscopic and submicroscopic changes in the coating-substrate system. The quantitative and qualitative data this provides must then be interpreted so that a prediction can be made as to whether the coating will fail, and if possible, why.

Macroscopic changes can be divided into two types:

1. Changes that can be seen by the unaided eye or with optical (light) microscopes, such as rust-through and creep from scribe
2. Large-scale changes that are found by measuring mechanical properties, of which the most important are adhesion to the substrate and the ability to prevent water transport

Changes in both the adhesion values obtained in before-and-after testing and in the failure loci can reveal quite a bit about aging and failure mechanisms. Changes in barrier properties, measured by electrochemical impedance spectroscopy (EIS), are important because the ability to hinder transport of electrolyte in solution is one of the more important corrosion-protection mechanisms of the coating.

One may be tempted to include such parameters as loss of gloss or color change as macroscopic changes. However, although these are reliable indicators of UV damage, they are not necessarily indicative of any weakening of the corrosion-protection ability of the coating system as a whole, because only the appearance of the topcoat is examined.

Submicroscopic changes cannot be seen with the naked eye or a normal laboratory light microscope but must instead be measured with advanced electrochemical or spectroscopic techniques. Examples include changes in chemical structure of the paint surface that can be found using Fourier transform infrared spectroscopy (FTIR) or changes in the morphology of the paint surface that can be found using atomic force microscopy (AFM). These changes can yield information about the coating-metal system, which is then used to predict failure, even if no macroscopic changes have yet taken place.

More sophisticated studies of the effects of aging factors on the coating include:

- Electrochemical monitoring techniques: AC impedance (EIS), Kelvin probe
- Changes in chemical structure of the paint surface using FTIR or x-ray photoelectron spectroscopy (XPS)
- Morphology of the paint surface using scanning electron microscopy (SEM) or AFM

8.2.1 GENERAL CORROSION

Direct evidence of corrosion can be obtained by macroscopic measurement of creep from scribe, rust intensity, blistering, cracking, and flaking.

8.2.1.1 Creep from Scribe

If a coating is properly applied to a well-prepared surface and allowed to cure, then general corrosion across the intact paint surface is not usually a major concern. However, once the coating is scratched and metal is exposed, the situation is dramatically different. The metal in the center of the scratch has the best access to oxygen and becomes cathodic. Anodes arise at the sides of the scratch, where paint, metal, and electrolyte meet [9]. Corrosion begins here and can spread outward from the scratch under the coating. The coating's ability to resist this spread of corrosion is a major concern.

Corrosion that begins in a scratch and spreads under the paint is called *creep* or *undercutting*. Creep is surprisingly difficult to quantify, because it is seldom uniform. Several methods are acceptable for measuring it, for example:

- Maximum one-way creep (probably the most common method), which is used in several standards, such as ASTM S1654
- Summation of creep at ten evenly spaced sites along the scribe
- Average two-way creep

None of these methods is satisfactory for describing filiform corrosion. The maximum one-way creep and the average two-way creep methods allow measurement of two values: general creep and filiform creep.

8.2.1.2 Other General Corrosion

Blistering, rust intensity, cracking, and flaking are judged in accordance with the standard ISO 4628 or the comparable standard ASTM D610. In these methods, the samples to be evaluated are compared to a set of standard photographs showing various degrees of each type of failure.

For face blistering, the pictures in the ISO standard represent blister densities from 2 to 5, with 5 being the highest density. Blister size is also numbered from 2 to 5, with 5 indicating the largest blister. Results are reported as blister density followed in parentheses by blister size (e.g., 4(S2) means blister density = 4 and blister size = 2); this is a way to quantify the result, "many small blisters."

For degree of rusting, the response of interest is rust under the paint, or rust bleed-through. Areas of the paint that are merely discolored on the surface by rusty runoff are not counted if the paint underneath is intact. The scale used by ISO 4628 in assigning degrees of rusting is shown in Table 8.1 [10].

Although the ASTM and ISO standards are comparable in methodology, their grading scales run in opposite directions. In measuring rust intensity or blistering,

TABLE 8.1
Degrees of Rusting

Degree	Area Rusted (%)
Ri 0	0
Ri 1	0.05
Ri 2	0.5
Ri 3	1
Ri 4	8
Ri 5	40–50

Source: ISO 4628/3-1982, *Designation of degree of rusting*, International Organization for Standardization, Geneva, 1982.

the ASTM standard uses 10 for defect-free paint and 0 for complete failure. The ISO standard uses 0 for no defects and the highest score for complete failure. These standards have faced some criticisms, mainly the following:

• They are too subjective.
• They assume an even pattern of corrosion over the surface.

Proposals have been made to counter the subjective nature of the tests by, for example, adding grids to the test area and counting each square that has a defect. The assumption of an even pattern of corrosion is questioned on the grounds that corrosion, although severe, can be limited to one region of the sample. Systems have been proposed to more accurately reflect these situations, for example, reporting the percentage of the surface that has corrosion and then grading the corrosion level within the affected (corroded) areas. For more information on this, the reader is directed to Appleman's review [2].

8.2.2 ADHESION

Many methods are used to measure adhesion of a coating to a substrate. The most commonly used methods belong to one of the following two groups: direct pull-off methods (e.g., ISO 4624) or cross-cut methods (e.g., ISO 2409). The test method must be specified; details of pull-stub geometry and adhesive used in direct pull-off methods are important for comparing results and must be reported.

8.2.2.1 The Difficulty of Measuring Adhesion

It is impossible to mechanically separate two well-adhering bodies without deforming them; the fracture energy used to separate them is therefore a function of both the interfacial processes and bulk processes within the materials [11]. In polymers, these bulk processes are commonly a complex blend of plastic and elastic deformation modes and can vary greatly across the interface. This leads to an interesting conundrum: the fundamental understanding of the wetting of a substrate by a liquid coating, and the subsequent adhesion of the cured coating to the substrate is one of the best-developed areas of coatings science, yet methods for the practical measurement of adhesion are comparatively crude and unsophisticated.

It has been shown that experimentally measured adhesion strengths consist of basic adhesion plus contributions from extraneous sources. Basic adhesion is the adhesion that results from intermolecular interactions between the coating and the substrate; extraneous contributions include internal stresses in the coating and defects or extraneous processes introduced in the coating as a result of the measurement technique itself [11]. To complicate matters, the latter can decrease basic adhesion by introducing new, unmeasured stresses or can increase the basic adhesion by relieving preexisting internal stresses.

The most commonly used methods of detaching coatings are applying a normal force at the interface plane or applying lateral stresses.

8.2.2.2 Direct Pull-off Methods

Direct pull-off (DPO) methods measure the force-per-unit area necessary to detach two materials, or the work done (or energy expended) in doing so. DPO methods employ normal forces at the coating-substrate interface plane. The basic principle is to attach a pulling device (a stub or dolly) to the coating by glue, usually cyanoacrylates, and then to apply a force to it in a direction perpendicular to the painted surface, until either the paint pulls off the substrate or failure occurs within the paint layers (see Figure 8.1).

An intrinsic disadvantage of DPO methods is that failure occurs at the weakest part of the coating system. This can occur cohesively within a coating layer; adhesively between coating layers, especially if the glue has created a weak boundary layer within the coating; or adhesively between the primer layer and the metal

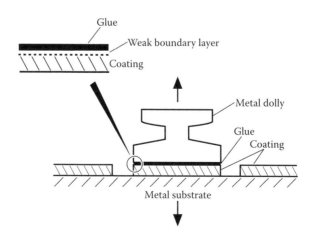

FIGURE 8.1 Direct pull-off adhesion measurement.

substrate, depending on which is the weakest link in the system. Therefore, adhesion of the primer to the metal is not necessarily what this method measures, unless it is at this interface that the adhesion is the weakest.

DPO methods suffer from some additional disadvantages:

- Tensile tests usually involve a complex mixture of tensile and shear forces just before the break, making interpretation difficult.
- Stresses produced in the paint layer during setting of the adhesive may affect the values measured (a glue/paint interactions problem).
- Nonuniform tensile load distributions over the contact area during the pulling process may occur. Stress concentrated in a portion of the contact area leads to failure at these points at lower values than would be seen under even distribution of the load. This problem usually arises from the design of the pulling head.

Unlike lateral stress methods, DPO methods can be used on hard or soft coatings. As previously mentioned, however, for a well-adhering paint, these methods tend to measure the cohesive strength of the coating, rather than its adhesion to the substrate.

With DPO methods, examination of the ruptured surface is possible, not only for the substrate but also for the test dolly. A point-by-point comparison of substrate and dolly surfaces makes it possible to fairly accurately determine interfacial and cohesive failure modes.

8.2.2.3 Lateral Stress Methods

Methods employing lateral stresses to detach a coating include bend or impact tests and scribing the coating with a knife, as in the cross-cut test.

In the cross-cut test, which is the most commonly used of the lateral stress methods, knife blades scribe the coating down to the metal in a grid pattern. The spacing of the cuts is usually determined by the coating thickness. Standard guidelines are given in Table 8.2. The amount of paint removed from the areas adjacent to, but not touched by, the blades is taken as a measurement of adhesion. A standard scale for evaluation of the amount of flaking is shown in Table 8.3.

Analysis of the forces involved is complex because both shear and peel can occur in the coating. The amount of shearing and peeling forces created at the knife

---

TABLE 8.2
Spacing of Cuts in Cross-Cut Adhesion

Coating thickness	Spacing of the cuts
Less than 60 μm	1 mm
60 μm–120 μm	2 mm
Greater than 120 μm	3 mm

TABLE 8.3
Evaluation of the Amount of Flaking

Grade	Description
0	Very sharp cuts. No material has flaked.
1	Somewhat uneven cuts. Detachment of small flakes of the coating at the intersections of the cuts.
2	Clearly uneven cuts. The coating has flaked along the edges and at the intersections of the cuts.
3	Very uneven cuts. The coating has flaked along the edges of the cuts partly or wholly in large ribbons and it has flaked partly or wholly on different parts of the squares. A cross-cut area of no more than 35% may be affected.
4	Severe flaking of material. The coating has flaked along the edges of the cuts in large ribbons and some squares have been detached partly or wholly. A cross-cut area of no more than 65% may be affected.
5	A cross-cut area greater than 65% is affected.

tip depends not only on the energy with which the cuts are made (i.e., force and speed of scribing) but also on the mechanical properties — plastic versus elastic deformation — of the coating. For example, immediately in front of the knife-edge, the upper surface of the paint undergoes plastic deformation. This deformation produces a shear force down at the coating-metal interface, underneath the rim of indentation in front of the knife-edge [11].

A major drawback to methods using lateral stresses is that they are extremely dependent on the mechanical properties of the coating, especially how much plastic and elastic deformation the coating undergoes. Paul has noted that many of these tests result in cohesive cracking of coatings [11]. Coatings with mostly elastic deformation commonly develop systems of cracks parallel to the metal-coating interface, leading to flaking at the scribe and poor test results. Coatings with a high proportion of plastic deformation, on the other hand, perform well in this test — even though they may have much poorer adhesion to the metal substrate than do hard coatings.

Elastic deformation means that little or no rounding of the material occurs at the crack tip during scribing. Almost all the energy goes into crack propagation. As the knife blade moves, more cracks in the coating are initiated further down the scribe. These propagate until two or more cracks meet and lead to flaking along the scribe. The test results can be misleading; epoxies, for example, usually perform worse than softer alkyds in cross-cut testing, even though, in general, they have much stronger adhesion to metal.

For very hard coatings, scribing down to the metal may not be possible. Use of the cross-cut test appears to be limited to comparatively soft coatings. Because the test is very dependent on deformation properties of the coatings, comparing cross-cut results of different coatings to each other is of questionable value. However, the test may have some value in comparing the adhesion of a single coating to various substrates or pretreatments.

8.2.2.4 Important Aspects of Adhesion

The failure loci — where the failure occurred — can yield very important information about coating weaknesses and eventual failures. Changes in failure loci related to aging of a sample are especially revealing about what is taking place within and under a coating system.

Adhesion measurements are performed to gain information regarding the mechanical strengths of the coating-substrate bonds and the deterioration of these bonds when the coatings undergo environmental stresses. A great deal of work has been done to develop better methods for measuring the strengths of the initial coating-substrate bonds.

By comparison, little attention has been given to using adhesion tests to obtain information about the mechanism of deterioration of either the coating or its adhesion to the metal. This area deserves greater attention because studying the failure loci in adhesion tests before and after weathering can yield a great deal of information about why coatings fail.

Finally, it is important to remember that adhesion is only one aspect of corrosion protection. At least one study shows that the coating with the best adhesion to the metal did not provide the best corrosion protection [12]. Also, studies have found that there is no obvious relationship between initial adhesion and wet adhesion [13].

8.2.3 BARRIER PROPERTIES

Coatings, being polymer-based, are naturally highly resistant to the flow of electricity. This fact is utilized to measure water uptake by and transport through the coating. The coating itself does not conduct electricity; any current passing through it is carried by electrolytes in the coating. Measuring the electrical properties of the coating makes it possible to calculate the amount of water present (called *water content* or *solubility*) and how quickly it moves (called *diffusion coefficient*). The technique used to do this is EIS.

An intact coating is described in EIS as a general equivalent electrical circuit, also known as the *Randles model* (see Figure 8.2). As the coatings become more porous or local defects occur, the model becomes more complex (see Figure 8.3).

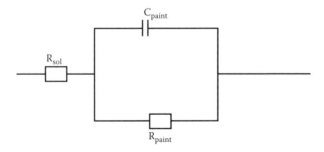

FIGURE 8.2 Equivalent electric circuit to describe an intact coating. R_{sol} is the solution resistance, C_{paint} is the capacitance of the paint layer, and R_{paint} is the resistance of the paint layer.

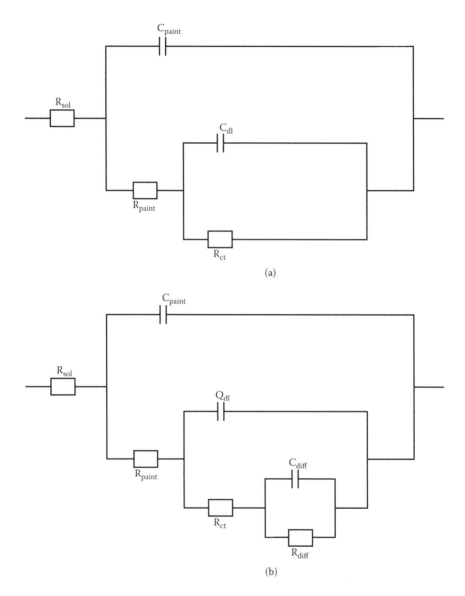

FIGURE 8.3 Equivalent electric circuits to describe a defective coating. C_{dl} is the double layer capacitance, R_{ct} is the charge transfer resistance of the corrosion process, Q_{dl} is the constant phase element, C_{diff} is the diffuse layer capacitance, and R_{diff} is the diffuse layer resistance.

Circuit A in Figure 8.3 is the more commonly used model; it is sometimes referred to as the *extended Randles model* [12, 14, 15].

EIS is an extremely useful technique in evaluating the ability of a coating to protect the underlying metal. It is frequently used as a "before-and-after" test because it is used to compare the water content and diffusion coefficient of the coating before and after aging (accelerated or natural exposure). Krolikowska [16] has suggested that for a coating to provide corrosion protection to steel, it should have an initial impedance of at least 10^8 /cm^2, a value also suggested by others [15], and that after aging, the impedance should have decreased by no more than three orders of magnitude. Sekine has reported blistering when the coating resistance falls below 10^6 /cm^2, regardless of coating thickness [17-19].

For more in-depth reviews of the fundamental concepts and models used in EIS to predict coating performance, the reader is directed to the research of Kendig and Scully [20] and Walter [21-23].

8.2.4 SCANNING KELVIN PROBE

The scanning Kelvin probe (SKP) provides a measure of the Volta potential (work function) that is related to the corrosion potential of the metal, without touching the corroding surface [24]. The technique can give a corrosion potential distribution, with a spatial resolution of 50 to 100 μm, below highly isolating polymer films. The SKP is an excellent research tool to study the initiation of corrosion at the metal/polymer interface.

Figures 8.4 and 8.5 show the Volta potential distribution for a coil-coated sample before and after 5 weeks of weakly accelerated field testing [25]. In the "after"

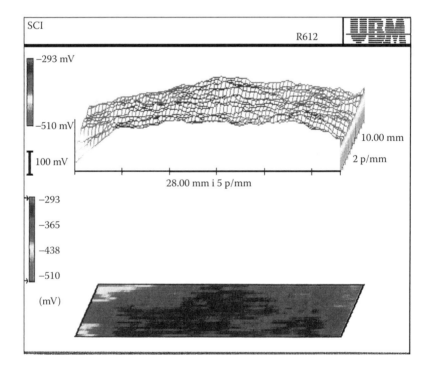

FIGURE 8.4 Volta distribution (mV) of coated steel before exposure.
Source: Forsgren, A. and Thierry, D., SCI Rapport 2001:4E. Swedish Corrosion Institute (SCI), Stockholm, 2001. Photo courtesy of SCI.

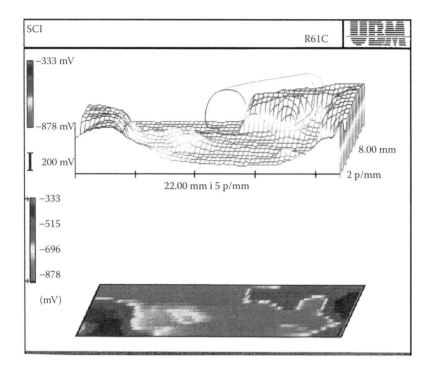

FIGURE 8.5 Volta distribution (mV) of coated steel before (top) and after (bottom) 5 weeks' weakly accelerated field exposure.
Source: Forsgren, A. and Thierry, D., SCI Rapport 2001:4E. Swedish Corrosion Institute (SCI), Stockholm, 2001. Photo courtesy of SCI.

figure, a large zone at low potential (−850 to −750 mV/NHE) can be clearly seen. Delamination, corrosion, or both is occurring at the transition area between the "intact" metal/polymer interface (zones at higher potential values, −350 to −200 mV/NHE) and more negative electrode potentials. The corrosion that is starting here after 5 weeks will not be visible as blisters for nearly 2 years at the Bohus-Malmön coastal station in Sweden [25].

8.2.5 Scanning Vibrating Electrode Technique

The scanning vibrating electrode technique (SVET) is used to quantify and map localized corrosion. The instrument moves a vibrating probe just above (100 μm or less) the sample surface, measuring and mapping the electric fields that are generated in the adjacent electrolyte as a result of localized electrochemical or corrosion activity. It is a well-established tool in researching localized events, such as pitting corrosion, intergranular corrosion, and coating defects. The SVET, which gives a two-dimensional distribution of current, is similar in many respects to the SKP; in fact, some instrument manufacturers offer a combined SVET/SKP system.

8.2.6 ADVANCED ANALYTICAL TECHNIQUES

For the research scientist or the well-equipped failure analysis laboratory, several advanced analytical techniques can prove useful in studying protective coatings. Many such techniques are based on detecting charged particles that come from, or interact with, the surface in question. These require high (10^{-5} or 10^{-7} torr) or ultrahigh vacuum (less than 10^{-8} torr), which means that samples cannot be studied in situ [26].

8.2.6.1 Scanning Electron Microscopy

Unlike optical microscopes, SEM does not use light to examine a surface. Instead, SEM sends a beam of electrons over the surface to be studied. These electrons interact with the sample to produce various signals: x-rays, back-scattered electrons, secondary electron emissions, and cathode luminescence. Each of these signals has slightly different characteristics when they are detected and photographed. SEM has very high depth of focus, which makes it a powerful tool for studying the contours of surfaces.

Electron microscopes used to be found only in research institutes and more sophisticated industrial laboratories. They have now become more ubiquitous; in fact, they are an indispensable tool in advanced failure analysis and are found in most any laboratory dealing with material sciences.

8.2.6.2 Atomic Force Microscopy

AFM provides information about the morphology of a surface. Three-dimensional maps of the surface are generated, and some information of the relative hardness of areas on the surface can be obtained. AFM has several variants for different sample surfaces, including contact mode, tapping mode, and phase contrast AFM. Soft polymer surfaces, such as those found in many coatings, tend to utilize tapping mode AFM.

In waterborne paint research, AFM has proven an excellent tool for studying coalescence of latex coatings [27-30]. It has also been used to study the initial effect of waterborne coatings on steel before film formation can occur, as shown in Figures 8.6 and 8.7 [31].

8.2.6.3 Infrared Spectroscopy

Infrared spectroscopy is a family of techniques that can be used to identify chemical bonds. When improved by Fourier transform mathematical techniques, the resulting test is known as FTIR. An FTIR scan can be used to identify compounds rather in the same way as fingerprints are used to identify humans: an FTIR scan of the sample is compared to the FTIR scans of "known" compounds. If a positive match is found, the sample has been identified; an example is shown in Figure 8.8. Not surprisingly, FTIR results are sometimes called "fingerprints" by analytical chemists.

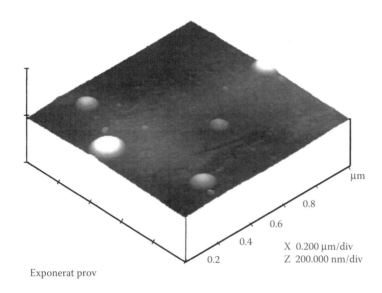

X 0.200 µm/div
Z 200.000 nm/div

Exponerat prov

FIGURE 8.6 Example of AFM imaging.
Photo courtesy of SCI.

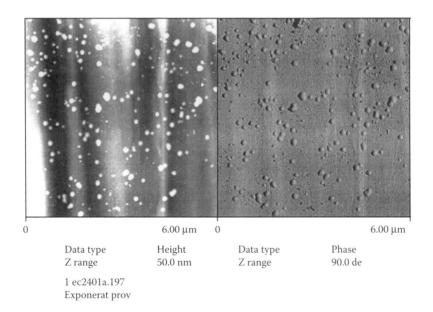

Data type	Height	Data type	Phase
Z range	50.0 nm	Z range	90.0 de

1 ec2401a.197
Exponerat prov

FIGURE 8.7 Example of AFM imaging.
Photo courtesy of SCI.

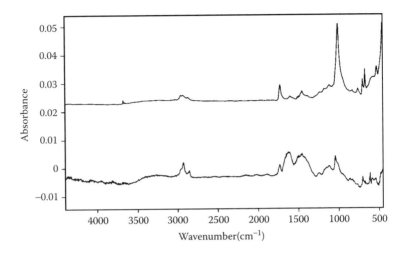

FIGURE 8.8 Example of FTIR fingerprinting.
Photo courtesy of SCI.

The most important FTIR techniques include:

• Attenuated total reflectance (ATR), in which a sample is placed in close
  contact with the ATR crystal. ATR is excellent on smooth surfaces that
  do not degrade during the test.
• Diffuse internal reflectance (DRIFT). DRIFT uses potassium bromide
  pellets for sample preparation and, therefore, has certain limitations in
  use with hygroscopic materials.
• Photoacoustic spectroscopy (PAS). In PAS, the sample surface absorbs
  radiation, heats up, and gives rise to thermal waves. These cause pressure
  variations in the surrounding gas, which are transmitted to a microphone
  — hence the acoustic signal [32].

8.2.6.4 Electron Spectroscopy

Electron spectroscopy is a type of chemical analysis in which a surface is bombarded
with particles or irradiated with photons so that electrons are emitted from it
[26, 33]. Broadly speaking, different elements emit electrons in slightly different
ways; so an analysis of the patterns of electrons emitted — in particular, the kinetic
energy of the electron in the spectrometer and the energy required to knock it off
the atom (binding energy) — can help identify the atoms present in the sample.

There are several types of electron spectroscopy techniques, each differing in
their irradiation sources. The one most important to coatings research, XPS (or
electron spectroscopy for chemical analysis [ESCA]), uses monochromatic x-rays.
XPS can identify elements (except hydrogen and helium) located in the top 1 to 2 Nm
of a surface [2, 26, 33]. It can also yield some information about oxidation states
because the binding energy of an electron is somewhat affected by the atoms around

it. It can be a powerful research tool and has been used, for example, to characterize the anodic oxide films on platinum that resulted from different anodizing methods [34]. It is also extremely useful for confirming theories of mechanisms in cases where the presence or absence of one or more elements is significant.

8.2.6.5 Electrochemical Noise Measurement

Electrochemical noise measurement (ENM) has attracted attention since it was first applied to anticorrosion coatings in the late 1980s [35, 36]. The noise consists of fluctuations in the current or potential that occur during the course of corrosion. The underlying idea is that these fluctuations in current or potential are not entirely random. An unavoidable minimum noise associated with current flow will always be random. However, if this minimum can be predicted for an electrochemical reaction, then analysis of the remainder of the noise may yield information about other processes, such as pitting corrosion, mass transport fluctuations, and the formation of bubbles (i.e., hydrogen formed at the cathode).

The theoretical treatment of electrochemical noise is not complete. There does not yet seem to be consensus on which signal analysis techniques are most useful. It is fairly clear, however, that understanding of ENM requires a good working knowledge of statistics; anyone setting out to master the technique must steel themselves to hear of kurtosis, skewness, and block averages rather frequently.

In the future, this technique may become a standard research tool for localized corrosion processes that give strong electrochemical noise signals, such as microbial corrosion and pitting corrosion.

8.3 CALCULATING AMOUNT OF ACCELERATION AND CORRELATIONS

Accelerated tests are most commonly used in one of two ways:

1. To compare or rank a series of samples in order to screen out unsuitable coatings or substrates (or conversely, in order to find the most applicable ones)
2. To predict whether a coating/substrate combination will give satisfactory performance in the field — and for how long

This requires that it be possible to calculate both the amount of acceleration the test causes and how uniform this amount of acceleration is over a range of substrates and coatings.

In order to be useful in comparing different coating systems or substrates, an accelerated test must cause even acceleration of the corrosion process among all the samples being tested. Different paint types have different corrosion-protection mechanisms; therefore, accentuating one or more stresses — such as heat or wet time — can be expected to produce different amounts of acceleration of corrosion among a group of coatings. The same holds true for substrates. As the stress or stresses are further accentuated — higher temperatures, more wet time, more salt, more UV light

— the variation in the corrosion rates for different coatings or different substrates increases. Three samples sitting side by side in an accelerated test, for example, may have a 3X, a 2X, and an 8X acceleration rates due to different vulnerabilities in different coatings. The problem is that the person performing the test, of course, does not know the acceleration rate for each sample. This can lead to incorrect ranking of coatings or substrates when the accelerated test is completed.

The problem for any acceleration method, therefore, is to balance the amount of acceleration obtained, with the variation (among different coatings or substrates). The variation should be minimal and the acceleration should be maximal; this is not trivial to evaluate because, in general, a higher acceleration can be expected to produce more variation in acceleration rate for the group of samples.

8.3.1 ACCELERATION RATES

The amount of acceleration provided by a laboratory test could be considered as quite simply the ratio of the amount of corrosion seen in the laboratory test to the amount seen in field exposure (also known as "reference") over a comparable time span. It is usually reported as 2X, 10X, and so on, where 2X would be corrosion in the lab occurring twice as quickly as in the field, as shown here:

$$A = \frac{X_{accel}}{X_{field}} \cdot \frac{t_{field}}{t_{accel}}$$

Where:
A is the rate of acceleration
X_{accel} is the response (creep from scribe) from the accelerated test
X_{field} is the response from field exposure
t_{accel} is the duration of the acceleration test
t_{field} is the duration of the field exposure

For example, after running test XYZ in the lab for 5 weeks, 4 mm creep from scribe was seen on a certain sample. After 2 years' outdoor exposure, an identical sample showed 15 mm creep from scribe. The rate of acceleration, A, could be calculated as:

$$A = \frac{(4\,mm/5\,weeks)}{(15\,mm/104\,weeks)} = 5.5$$

8.3.2 CORRELATION COEFFICIENTS OR LINEAR REGRESSIONS

Correlation coefficients can be considered indicators of the uniformity of acceleration within a group of samples. Correlations by linear least square regression are calculated for data from samples run in an accelerated test versus the response of identical samples in a field exposure. A high correlation coefficient is taken as an indication that the test accelerates corrosion more or less to the same degree for all samples in the group. One drawback of correlation analyses that use least square regression is that they are sensitive to the distribution of data [37].

8.3.3 Mean Acceleration Ratios and Coefficient of Variation

Another interesting approach to evaluating field data versus accelerated data is the mean acceleration ratio and coefficient of variation [37].

To compare data from a field exposure to data from an accelerated test for a set of panels, the acceleration ratio *for each type of material* (i.e., coating and substrate) is calculated by dividing the average result from the accelerated test by the corresponding reference value, usually from field exposure. These results are then summed up for all the panels in the set and divided by the number of panels in the set to give the mean acceleration ratio. That is,

$$MVQ = \frac{\sum_{i=1}^{n} \frac{X_{i,accel}}{X_{i,field}}}{n} \ +/- \sigma_{n-1}$$

Where:
MVQ is the mean value of quotients
$X_{i,accel}$ is the response (creep from scribe) from the accelerated test for each sample i
$X_{i,field}$ is the response from field exposure for each sample i
n is the number of samples in the set [37, 38]

This is used to normalize the standard deviation by dividing it by the mean value (MVQ):

$$Coefficent \ of \ variation \ = \ \frac{\sigma_{n-1}}{MVQ}$$

$$Test \ accel. \ = \ MVQ \cdot \frac{t_{field}}{t_{accel}}$$

The coefficient of variation combines the amount of acceleration provided by the test with how uniformly the corrosion is accelerated for a set of samples. It is desirable, of course, for an acceleration test to accelerate the corrosion rate more or less uniformly for all the samples; that is, the standard deviation should be as low as possible. It follows naturally that the ratio of deviation to mean acceleration should be as close to 0 as possible. A high coefficient of variation means that, for each set of data, there is more spread in the amount of acceleration than there is actual acceleration.

8.4 SALT SPRAY TEST

The salt spray (fog) test ASTM B117 ("*Standard Practice for Operating Salt Spray (Fog) Testing Apparatus*") is one of the oldest corrosion tests still in use. Despite a widespread belief among experts that the salt spray test is of no value in predicting performance, or even relative ranking, of coatings in most applications, it is the most frequently specified test for evaluating paints and substrates.

8.4.1 THE REPUTATION OF THE SALT SPRAY TEST

The salt spray test has such a poor reputation among workers in the field that the word "infamous" is sometimes used as a prefix to the test number. In fact, nearly every peer-reviewed paper published these days on the subject of accelerated testing starts with a condemnation of the salt spray test [39-44]. For example:

- "In fact, it has been recognized for many years that when ranking the performance levels of organic coating systems, there is little if any correlation between results from standard salt spray tests and practical experience." [3]
- "The well-known ASTM B117 salt spray test provides a comparison of cold-rolled and electrogalvanized steel within several hundred hours. Unfortunately, the salt spray test is unable to predict the well-known superior corrosion resistance of galvanized relative to uncoated cold rolled steel sheet." [45]
- "Salt spray provides rapid degradation but has shown poor correlation with outdoor exposures; it often produces degradation by mechanisms different from those seen outdoors and has relatively poor precision." [46]

Many studies comparing salt spray results and actual field exposure have been performed. Coating types, substrates, locations, and length of time have been varied. No correlations have been found to exist between the salt spray and the following service environments:

- Galveston Island, Texas (16 months), 800 meters from the sea [47]
- Sea Isle City, New Jersey (28 months), a marine exposure site [48]
- Daytona Beach, Florida (3 years) [49]
- Pulp mills at Lessebo and Skutskar, Sweden, painted hot-rolled steel substrates (4 years) [50] and painted aluminium, galvanized steel and carbon steel substrates (5 years) [51]
- Kure Beach, North Carolina, a marine exposure site [52-54]

8.4.2 SPECIFIC PROBLEMS WITH THE SALT SPRAY TEST

Appleman and Campbell [55] have examined each of the accelerating stresses in the salt spray test and its effect on the corrosion mechanism compared to outdoor or "real-life" exposure. They found the following flaws in the salt spray test:

a. Constant humid surface
 - Neither the paint nor the substrate experience wet /dry cycles. Corrosion mechanisms may not match those seen in the field; for example, in zinc-rich coatings or galvanized substrates, the zinc is not likely to form a passive film as it does in the field.

- Water uptake and hydrolysis are greater than in the field.
- A constant water film with high conductivity is present, which does not happen in the field.

b. Elevated temperature
- Water, oxygen, and ion transport are greater than in the field.
- For some paints, the elevated temperature of the test comes close to the glass transition temperature of the binder.

c. **High chloride concentration** (effect on corrosion depends on the type of protection the coating offers)

- For sacrificial coatings, such as zinc-rich primers, the high chloride content together with the constant high humidity means that the zinc is not likely to form a passive film as it does in the field.
- For inhibitive coatings, chlorides adsorb on the metal surface, where they prevent passivation.
- For barrier coatings, the osmotic forces are much less than in the field; in fact, they may be reversed completely from that which is seen in reality. In the salt spray test, corrosion at a scribe or defect is exaggeratedly aggressive compared with a scribe under intact paint.

Lyon, Thompson, and Johnson [56] point out that the high sodium chloride content of the salt spray test can result in corrosion morphologies and behaviors that are not representative of natural conditions. Harrison has pointed out that the test is inappropriate for use on zinc — galvanized substrates or primers with zinc phosphate pigments, for example — because, in the constant wetness of the salt spray test, zinc undergoes a corrosion mechanism that it would not undergo in real service [57]. This is a well-known and well-documented phenomenon and is discussed in depth in chapter 7.

8.4.3 Importance of Wet/Dry Cycling

Skerry, Alavi, and Lindgren have identified three factors of importance in the degradation and corrosion of painted steel that are not modeled by the salt spray test: wet/dry cycling, a suitable choice of electrolyte, and the effects of UV radiation (critical because of the breakdown of polymer bonds in the paint) [3].

Lyon, Thompson, and Johnson explain why wet/dry cycles are an important factor in an accelerated test method [56]:

Many studies have shown the specific importance of wetting and drying on atmospheric corrosion... On a dry metal surface, as the relative humidity (RH) is increased, the corrosion rate initially rises, then decreases to a relatively constant value which becomes greater as the RH is increased. A similar effect is observed during physical wetting and drying of a surface. Thus, on initial wetting, the corrosion rate rises rapidly as accumulated surface salts first dissolve. The rate then decreases as the surface electrolyte dilutes with continued wetting. The corrosion rate also rises significantly during drying because of both the increasing ionic activity as the surface electrolyte concentrates and the reduced diffusion layer thickness for oxygen as the condensed phase become thinner. However, eventually, when the ionic strength of the electrolyte layer becomes very high and salts begin to crystallize, the corrosion rate decreases.

Simpson, Ray, and Skerry agree that "cyclic wetting and drying of electrolyte layers from the panel surface is thought to stress the coating in a more realistic manner than, for example, a continuous ASTM B-117 salt spray test, where panels are placed in a constant, high relative humidity (RH) environment" [58]. Several workers in this field have reported that cyclic testing with a significant amount of drying time yields more realistic results on zinc-coated substrates [59-61].

REFERENCES

1. Goldie, B., *Prot. Coat. Eur.*, 1, 23, 1996.
2. Appelman, B., *J. Coat. Technol.*, 62, 57, 1990.
3. Skerry, B.S., Alavi, A., and Lindgren, K.I. *J. Coat. Technol.*, 60, 97, 1988.
4. Townsend, H., Development of an improved laboratory corrosion test by the automotive and steel industries, in *Proc. 4th Annual ESD Advanced Coatings Conference*, Dearborn, MI, 8-10 November 1994. Engineering Society of Detroit (ESD), 1994.
5. *Prfung des korrosionsschutzes von kraftfahrzeuglackierungen bei zyklisch wechselnder beanspruchung*, Std. VDA 621-415, German Association of the Automotive Industry, Frankfurt, Germany.
6. *Accelerated corrosion test*, Corporate Standard STD 423-0014, Issue 1, Volvo Group, Gothenburg, 2003.
7. Ström, M., in *Proc. Conf. Automotive Corrosion and Prevention*. Dearborn, MI, Dec 4-6, 1989, Society of Automotive Engineers, Warrendale, PA, 1989.
8. Ström, M. and Ström, G., *Proc. Skan Zink '91*, Helsingör (Sweden), Sept. 1991.
9. Pletcher, D., *A First Course in Electrode Processes*, The Electrochemical Consultancy Ltd., Romsey, England, 1991, 229.
10. ISO 4628/3-1982-*Designation of degree of rusting*, International Organization for Standardization, Geneva, 1982.
11. Paul, S., (Ed.), *Surface Coatings: Science & Technology,* 2nd ed., John Wiley & Sons, Chichester, England, 1996.
12. Sacco, E.A. et al., *Lat. Am. Appl. Res.* 32, 4, 2002.
13. Walker, P., *Paint Technol.*, 31, 22. 1967.
14. Özcan, M., Dehri, I. and Erbil, M., *Prog. Org. Coat.*, 44, 279, 2002.
15. Lavaert, V. et al., *Prog. Org. Coat.*, 38, 213, 2000.
16. Krolikowska, A., *Prog. Org. Coat.*, 39, 37, 2000.
17. Sekine, I., *Proc. 99th Symposium on Corrosion Protection*, Tokyo, Japan, 1994, 51.
18. Sekine, I. and Yuasa, M., *Proc. Annual Meeting of the Japan Society of Colour Material*, Tokyo, 1995, 70.
19. Sekine, I., *Prog. Org. Coat.*, 31, 73, 1997.
20. Kendig, M. and Scully, J. *Corrosion*, 46, 22, 1990.
21. Walter, G.W., *Corros. Sci.*, 32, 1041, 1991.
22. Walter, G.W., *Corros. Sci.*,32, 1059, 1991.
23. Walter, G.W., *Corros. Sci.*,32, 1085, 1991.
24. M. Stratmann et al., *Corros. Sci.*,6, 715, 1990.
25. Forsgren, A. and Thierry, D., *Corrosion properties of coil-coated galvanized steel, using field exposure and advanced electrochemical techniques*, SCI Rapport 2001:4E. Swedish Corrosion Institute (SCI), Stockholm, 2001.

26. Bard, A.J. and Faulkner, L.R., *Electrochemical Methods*, John Wiley & Sons, New York, 1980, chap. 6.

27. Glicinski, A.G. and Hegedus, C.R., *Prog. Org. Coat.*, 32, 81, 1997.

28. Joanicot, M., Granier, V. and Wong, K., *Prog. Org. Coat.*, 32, 109, 1997.

29. Gerharz, B., et al., *Prog. Org. Coat.*, 32, 75, 1997.

30. Tzitzinou, A. et al., *Prog. Org. Coat.*, 35, 89, 1999.

31. Forsgren, A. and Persson, D., Changes in the Surface Energy of Steel Caused by Acrylic Waterborne Paints Prior to Cure, SCI Rapport 2000:5E. Swedish Corrosion Institute (SCI), Stockholm, 2000.

32. Almeida, E., Balmayore, M. and Santos, T., *Prog. Org. Coat.*, 44, 233, 2002.

33. Shaw, D., *Introduction to Colloid and Surface Chemistry*, 4th ed., Butterworth-Heinemann Ltd., 1991, chapters 3 and 5.

34. Kim, K.S, Winograd, N. and Davis, R.E., *J. Amer. Chem. Soc.*, 93, 6296, 1971.

35. Skerry, B.S. and Eden, D.A., *Prog. Org. Coat.*, 15, 269, 1987.

36. Chen, C.T. and Skerry, B.S., *Corrosion*, 47, 598, 1991.

37. Ström, M. and Ström. G., *SAE Technical Paper Series, 932338*. Society of Automotive Engineers, Warrendale, PA, 1993.

38. Ström, M., Utviklingen av metallbelegg i bilindustrin: Status, trender og Volvos erfaringer, in *Proc Overflatedager '92*, Trondheim, 1992. (in Norwegian)

39. LaQue, F.L., *Marine Corrosion*, Wiley, New York, 1975.

40. Lambert, M.R., et al., *Ind. Eng. Chem. Prod. Res. Dev.*, 24, 378, 1985.

41. Lyon, S.B. et al., *Corrosion*, 43, 12, 1987.

42. Timmins, F.D., *J. Oil Color Chem. Assn.*, 62, 131, 1979.

43. Funke, W., *J. Oil Color Chem. Assn.*, 62, 63, 1979.

44. Skerry, B.S. and Simpson, C.H., *Corrosion*, 49, 663, 1993.

45. Townsend, H.E., Development of an improved laboratory corrosion test by the automotive and steel industries, in *Proc. 4th Annual ESD Advanced Coating Conference*, Dearborn, MI, 8-10 November 1994. Engineering Society of Detroit (ESD), 1994.

46. Appleman, B., *J. Prot. Coat. Linings*, 6, 71, 1989.

47. Struemph, D.J. and Hilko, J., *IEEE Trans. on Power Delivery*, PWRD-2, 823, 1987.

48. Chong, S.L., Comparison of laboratory testing method for bridge coatings, in *Proc. 4th World Congress on Coating Systems for Bridge and Steel Structures Bridging the Environment*, St. Louis, MO, 1995.

49. Rommal, H.E.G. et al., Accelerated test development for coil-coated steel building panels, in *Proc. Corrosion '98*, San Diego, CA, National Association of Corrosion Engineers (NACE), 1998, Paper 356.

50. Forsgren, A. and Palmgren, S., *Salt spray test vs. field results for coated samples: part I*, SCI Rapport 1998:4E, Swedish Corrosion Institute (SCI), Stockholm, 1998.

51. Forsgren, A., Rendahl, B. and Appelgren, C., *Salt spray test vs. field results for coated samples: part II*, SCI Rapport 1998:6E, Swedish Corrosion Institute (SCI), Stockholm, 1998.

52. Appleman, B.R., Bruno, J.A., and Weaver, R.E.F., *Performance of Alternate Coatings in the Environment (PACE) Volume I: Ten Year Field Data*, FHWA-RD-89-127, U.S. Federal Highway Administration, Washington D.C., 1989.

53. Appleman, B.R., Weaver, R.E.F. and Bruno, J.A., *Performance of Alternate Coatings in the Environment (PACE) Volume II: Five Year Field Data and Bridge Data of Improved Formulations*, FHWA-RD-89-235, U.S. Federal Highway Administration, Washington D.C., 1989.

54. Appleman, B.R., Weaver, R.E.F. and Bruno, J.A., *Performance of Alternate Coatings in the Environment (PACE) Volume III: Executive Summary*, FHWA-RD-89-236, U.S. Federal Highway Administration, Washington D.C., 1989.

55. Appleman, B.R. and Campbell, P.G., *J. Coat. Technol.*, 54, 17, 1982.

56. Lyon, S.B., Thompson, G.E. and Johnson, J.B., Materials evaluation using wet-dry mixed salt spray tests, in *New Methods for Corrosion Testing of Aluminum Alloys*, ASTM STP 1134, V.S. Agarwala and G.M. Ugiansky, Eds., American Society for Testing and Materials, Philadelphia, PA, 1992, 20.

57. Harrison, J.B. and Tickle, T.C., *J. Oil Color Chem. Assn.*, 45, 571, 1962.

58. Simpson, C.H, Ray, C.J. and Skerry, B.S., *J. Prot. Coat. Linings*, 8, 28, 1991.

59. Smith, D.M. and Whelan, G.W., *SAE Technical Paper Series, 870646*, Society of Automotive Engineers, Warrendale, PA, 1987.

60. Nowak, E.T., Franks, L.L. and Froman, G.W., *SAE Technical Paper Series, 820427*, Society of Automotive Engineers, Warrendale, PA, 1982.

61. Standish, J.V., Whelan, G.W. and Roberts, T.R., *SAE Technical Paper Series, 831810*. Society of Automotive Engineers, Warrendale, PA, 1983.

Index

Other Related Titles of Interest Include:

Corrosion Resistance Tables: Metals, Nonmetals, Coatings, Mortars, Plastics, Elastomers and Linings, and Fabrics, Fifth Edition (4 Volume Set)
Philip A. Schweitzer, P.E.
ISBN: 082475672X

Corrosion Mechanisms in Theory and Practice, Second Edition
Philippe Marcus
ISBN: 0824706668

Coatings Technology Handbook, Third Edition
Arthur A. Tracton
ISBN: 1574446495

Analytical Methods in Corrosion Science and Engineering
Philippe Marcus
ISBN: 0824759524

Paint and Coatings: Applications and Corrosion Resistance
Philip A. Schweitzer, P.E.
ISBN: 1574447025

And coming soon:

Corrosion Science Mechanisms, Mitigation and Monitoring
U. Kamachi Mudali and Baldev Raj
ISBN: 0849333741